THE MAGNETIC CIRCUIT

THE MAGNETIC CIRCUIT

Electromagnetic Engineering

V. Karapetoff

First Edition

ISBN 1-929148-37-2

WEXFORD COLLEGE PRESS
2003

PREFACE

THIS book, together with the companion book entitled " The Electric Circuit," is intended to give a student in electrical engineering the theoretical elements necessary for the correct understanding of the performance of dynamo-electric machinery, transformers, transmission lines, etc. The book also contains the essential numerical relations used in the predetermination of the performance and in the design of electrical machinery and apparatus. The whole treatment is based upon a very few fundamental facts and assumptions. The student must be taught to treat every electric machine as a particular combination of electric and magnetic circuits, and to base its performance upon the fundamental electromagnetic relations rather than upon a separate " theory " established for each kind of machinery, as is sometimes done.

The book is *not* intended for a beginner, but for a student who has had an elementary descriptive course in electrical engineering and some simple laboratory experiments. The treatment is somewhat different from that given in most other books dealing with magnetic phenomena. It is based directly upon the circuital relation, or interlinkage, between an electric current and the magnetic flux produced by it. This relation, and the law of induced electromotive force, are taken to be the fundamental phenomena of electro-magnetism. No use whatever is made of the usual artificial concepts of unit pole, magnetic charge, magnetic shell, etc. These concepts of mathematical physics, together with the law of inverse squares, embody the theory of action at a distance, and are both superfluous and misleading from the modern point of view of a continuous action in the medium itself.

The ampere-ohm system of units is used throughout, in accordance with Professor Giorgi's ideas, as is explained in the appendices. Those familiar with Oliver Heaviside's writings will notice his influence upon the author, in particular with regard to a uniform and rational nomenclature. The author trusts that his colleagues will judge his treatment and nomenclature upon their own merits, and not condemn them simply because they are different from the customary treatment.

In the first four chapters the student is introduced into the fundamental electromagnetic relations, and is made familiar with them by means of numerous illustrations taken from engineering practice. Chapters V to IX treat of the flux and magneto-motive force relations in electrical machinery, first at no load, and then under load when there is an armature reaction. The remaining four chapters are devoted to the phenomena of stored magnetic energy, namely inductance and tractive effort. The subject is treated entirely from the point of view of an electrical engineer, and the important relations and methods are illustrated by practical numerical problems, of which there are several hundred in the text. All matter of purely historical or theoretical interest has been left out, as well as special topics which are of interest to a professional designer only. An ambitious student will find a more exhaustive treatment in the numerous references given in the text.

Many thanks are due to the author's friend and colleague, Mr. John F. H. Douglas, instructor in electrical engineering in Sibley College, who read the manuscript and the proofs, checked the answers to the problems, and made many excellent suggestions for the text. Most of the sketches are original, and are the work of Mr. John T. Williams of the Department of Machine Design of Sibley College, to whom I am greatly indebted.

CORNELL UNIVERSITY, ITHACA, N. Y.,
          September, 1911.

CONTENTS

vii

viii CONTENTS

SUGGESTIONS TO TEACHERS

(1) THIS book is intended to be used as a text in a course which comprises lectures, recitations, computing periods, and home work. Purely descriptive matter has been omitted or only suggested, in order to allow the teacher more freedom in his lectures and to permit him to establish his own point of view. Some parts of the book are more suitable for recitations, others as reference in the computing room, others, again, as a basis for discussion in lectures or for brief theses.

(2) Different parts of the book are made as much as possible independent of one another, so that the teacher can schedule them as it suits him best. Moreover, most of the chapters are written according to the concentric method, so that it is not necessary to finish one chapter before starting on the next. One can thus cover the subject in an abridged manner, omitting the last parts of the chapters.

(3) The problems given at the end of nearly every article are an integral part of the book, and should, under no circumstances, be omitted. There is no royal way of obtaining a clear understanding of the underlying physical principles, and of acquiring an assurance in their practical application, except by the solution of numerical examples. It is convenient to assign each student the complete specifications of a machine of each kind, and ask him to solve the various problems in the text in application to these machines, in proportion as the book is covered. Numerous specifications and drawings of electrical machines will be found in the standard works of E. Arnold, H. M. Hobart, Pichelmayer and others, mentioned in the footnotes in the text. A first-hand acquaintance with these classical works on the part of the student is very desirable, however superficial this acquaintance may be.

(4) The book contains comparatively few sketches; this gives the student an opportunity to illustrate the important relations by sketches of his own. Making sketches and drawings of electric machines to scale, with their mechanical features, should be one of the important features of an advanced course, even though it may not be popular with some analytically-inclined students. Mechanical drawing develops precision of judgment, and gives the student a knowledge of machinery and apparatus that is tangible and concrete.

(5) The author has avoided giving definite numerical data, coefficients and standards, except in problems, where they are indispensable and where no general significance is ascribed to such data. His reasons are: (a) Numerical coefficients obscure the general exposition. (b) Sufficient numerical coefficients and design data will be found in good electrical hand-books and pocket-books, one of which ought to be used in conjunction with this text. (c) The student is liable to ascribe too much authority to a numerical value given in a text-book, while in reality many coefficients vary within wide limits, according to the conditions of a practical problem and with the progress of the art. (d) Most numerical coefficients are obtained in practice by assuming that the phenomenon in question occurs according to a definite law, and by substituting the available experimental data into the corresponding formula. This point of view is emphasized throughout the book, and gives the student the comforting feeling that he will be able to obtain the necessary numerical constants when confronted by a definite practical situation.

(6) The treatment of the magnetic circuit is made as much as possible analogous to that of the electrodynamic and electrostatic circuits treated in the companion book. The teacher will find it advisable to make his students perfectly fluent in the use of Ohm's law for ordinary electric circuits before starting on the magnetic circuit. The student should solve several numerical examples involving voltages and voltage gradients, currents and current densities, resistances, resistivities, conductances, and conductivities. He will then find very little difficulty in mastering the electrostatic circuit, and with these two the transition to the magnetic circuit is very simple indeed. The following table shows the analogous quantities in the three kinds of circuits.

Electrodynamic	Electrostatic	Magnetic

Voltage or e.m.f.
Voltage gradient (or
  electric intensity)
	Voltage or e.m.f.
	Voltage gradient (or
	  electric intensity)
		Magnetomotive force
		M.m.f. gradient (or
		  magnetic intensity)

Electric current Current density	Dielectric flux Dielectric flux density	Magnetic flux Magnetic flux density
Resistor Resistance Resistivity	Elastor Elastance Elastivity	Reluctor Reluctance Reluctivity
Conductor Conductance Conductivity	Permittor (condenser) Permittance (capacity) Permittivity (dielec- tric constant)	Permeator Permeance Permeability

LIST OF PRINCIPAL SYMBOLS

The following list comprises most of the symbols used in the text. Those not occurring here are explained where they appear. When, also, a symbol has a use different from that stated below, the correct meaning is given where the symbol occurs.

Symbol.	Meaning.	Page where defined or first used.
a	Air-gap	90
a	Width of commutator segment	237
A	Area	11
A_a	Area of flux per tooth pitch in the air	101
A_i	Area of flux per tooth pitch in the iron	101
(AC)	Number of ampere-conductors per centimeter	165
b	Thickness of transformer coil	211
b	Width of brush	236
b'	Thickness of mica	236
B	Flux density	14
B_m	Maximum value of the flux density	81
C_2	Number of secondary conductors	133
C_{pp}	Conductors per pole per phase	219
d	Duct width	94
e, E	Electromotive force	39, 65
f	Frequency	48
F	Mechanical force	274
F_t'	Tension per square centimeter	243
F_c'	Compression per square centimeter	244
H	Magnetic intensity	13
H_m	Maximum value of the magnetic intensity	81
i, I	Electric current	39, 205
i_λ	Magnetizing current	81
I_1	Current per armature branch	233
k, k'	Transformer constant	215, 221
k_a	Air-gap factor	90
k_b	Breadth factor	65
k_s	Slot factor	68
k_w	Winding-pitch factor	68
l	Length	11

NOTE: A careful distinction is made in the book between the expressions "centimeter cube" and "cubic centimeter." Centimeter cube means a cube having its side equal to one centimeter; one cubic centimeter is a volume, no matter what the shape of the body may be. We speak of reluctance per centimeter cube, and of magnetic energy per cubic centimeter.

THE MAGNETIC CIRCUIT

CHAPTER I

THE FUNDAMENTAL RELATION BETWEEN FLUX AND MAGNETOMOTIVE FORCE

1. A Simple Magnetic Circuit. The only known cause of magnetic phenomena is an electric current, or, more generally, electricity in motion.[1] The fundamental relation between an electric current and magnetism can be best studied with the simple arrangement shown in Fig. 1. A coil, CC, of very thin wire is uniformly wound in one layer on a spool made in the shape of a circular ring (toroid). The tubular space inside of the ring is filled with some "non-magnetic" material, so called; for instance, air, wood, etc. When a direct current is sent through the coil, the space inside the coil is found to be in a peculiar state, called the *magnetic state*. This magnetic state can be experimentally proved by various means, such as a compass needle, iron filings, etc. A region in which a magnetic state is manifested is called a *magnetic field*. Thus, in Fig. 1, the tubular space inside the coil is the magnetic field excited by the current in the coil CC.

No magnetic field is found in the space outside the coil upon exploring it with a magnetic needle or with iron filings. For reasons of symmetry, the field inside the ring is the same at all the cross-sections. Thus, a uniformly wound ring constitutes

---

[1] Werner v. Siemens, *Wiedemann's Annalen*, Vol. 24. (1885), p. 94; Larmor, *Ether and Matter* (1904), p. 108. The magnetism of a permanent magnet is probably due to molecular currents produced by some orbital motion of electrons within the atoms of iron. The older concepts of magnetic charges and free poles are summarily dismissed in this book as inadequate and artificial.

the simplest magnetic circuit, because the field is uniform and is entirely confined within the winding.

Iron filings orient themselves within the coil in directions indicated in Fig. 1 by the concentric lines with arrow-heads. These lines show that the medium is "magnetized" along circles concentric with the ring. Lines which show the directions in which a medium is magnetized are generally called *magnetic lines of force*.[1] They are analogous to the lines of electrostatic displacement, though their directions and physical nature are entirely different;

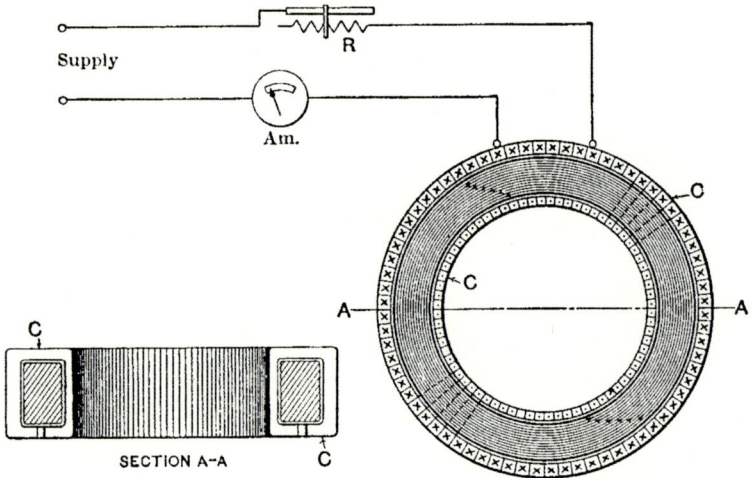

FIG. 1.—A simple magnetic circuit.

see Chapter XIV on the dielectric circuit, in the author's *Electric Circuit*.

The positive direction of the lines of force is purely conventional, and is defined as that in which the north-seeking end of a compass moves. Its relation to the current is, by experiment, that given by the right-hand screw rule. Namely, if the direction of the flow of a current is that of the rotation of a right-hand screw, the lines of force point in the direction of the progressive movement of the screw. Reversing the current reverses the direction

[1] For actual photographs, showing iron filings which map out the magnetic field inside of coils of various shapes, see Dr. Benischke, *Die Wissenschaftlichen Grundlagen der Elektrotechnik* (1907), p. 126; also his *Transformatoren* (1909), pp. 4, 6, and 57.

of the field; this fact can be demonstrated by a small compass needle. The positive direction of the lines of force is indicated in Fig. 1 by arrow heads. The direction of the current is shown in the conventional way by dots and crosses; namely, a dot indicates that the current is approaching the observer, while a cross indicates that the current is receding.

The magnetic state within the coil can also be explored by a small test-coil inserted into the field and connected to a galvanometer. When this coil is properly placed with respect to the field and then turned about its axis by some angle, the galvanometer shows a deflection, because a current is induced in the coil by the magnetic field. There are also other means for detecting a magnetic field, for which the reader is referred to books on physics.

The total magnetic field produced by a current is called a *magnetic circuit*, by analogy with the electric and the electrostatic circuits. Experiment shows that the magnetic lines of force are always closed curves like the stream lines of an electric current, or like the lines of electrostatic displacement (when these are completed through the conductors).

Fig. 1 exemplifies a fundamental law of electromagnetism; namely, an electric current creates a magnetic field in such directions that the lines of force are *linked* with the lines of flow of the current, in the same manner that the consecutive links of a chain are linked together. This law admits of no theoretical proof, and must be accepted as a fundamental experimental fact. Wherever there is an electric circuit there is also a magnetic field linking with it. The two are inseparable, and increase and decrease together. Each form of an electric circuit with a certain strength of current in it corresponds to a definite form of magnetic field. It is possible that the electric current and the magnetic field are but two different ways of looking upon one and the same phenomenon.

The linkages of magnetic lines with a current are seen more clearly in Fig. 11, which shows the magnetic field produced by a loop of wire, *aa*. It will be seen that the arrangement in Fig. 1 is more suitable for an elementary study, because the field is much more *uniform*, especially if the radial thickness of the ring is small as compared to its mean diameter, so that all the lines of force are of practically the same length.

The same right-hand screw rule applies in the case of Fig. 11

as in Fig. 1. When the current in the loop of wire circulates in the direction of rotation of a right-hand screw (toward the reader on the left), the lines of force *within the loop* point in the direction of the progressive movement of the screw (upward). The rule can be reversed by saying that when the direction of the lines of force around a wire is that of the rotation of a right-hand screw, the current in the wire flows in the direction of the progressive movement of the screw. The first statement is convenient in the case of a ring winding, the second in the case of a long straight conductor. Both rules can be combined into one by considering the exciting electric circuit and the resulting magnetic circuit as two consecutive links of a chain. When the arrow-head in one of the links (no matter which) points in the direction of rotation of a right-hand screw, the arrow-head in the other link, as it passes through the first, must point in the direction of the progressive movement of the screw.

2. Magnetomotive Force. Experiment shows that the magnetic field within the ring (Fig. 1) does not change if the current and the number of turns of the " exciting " winding vary so that their product remains the same. That is to say, 500 turns of wire with a current of 2 amperes flowing through each will produce the same field as 1000 turns with 1 ampere, or 200 turns with 5 amperes, because the product is equal to 1000 *ampere-turns* in all cases. Even one turn with 1000 amperes flowing through it will produce the same effect, provided that the turn is made of a wide sheet of metal spread over the whole surface of the ring, so as to make its action uniform throughout.

The reason for the above can be seen by considering 1000 separate turns with a current of 1 ampere flowing through each turn, and each turn supplied with current from an independent electrical source, say a dry cell. Connecting all the cells and all the turns in series gives 1000 turns with one ampere flowing through each. Connecting the cells and the turns in parallel results in one wide turn with 1000 amperes of current in it. Such changes in the electrical connections cannot affect the action of each current outside the wire, because the value of the current and the position of the turn is the same in both cases. Hence, the magnetic action depends only upon the number of turns each carrying 1 ampere, in other words, it depends upon the number of ampere-turns.

The number of ampere-turns of the exciting winding is called the *magnetomotive force* of the magnetic circuit, because these ampere-turns are the cause of the magnetic field. One ampere-turn is the logical unit of magnetomotive force. In the example above, the magnetomotive force is equal to 1000 ampere-turns. In electric machines the field excitation often reaches several thousand ampere-turns, and the magnetomotive force is for convenience sometimes measured in kiloampere-turns, one kilo-ampere-turn being equal to 1000 ampere-turns.

3. Magnetic Flux. The magnetic disturbance at each point within the ring has not only a direction, but also a magnitude. The disturbance is said to be in the form of a *flux*, for the following reason: One may think of the magnetic state as being due to the actual displacement of some hypothetical incompressible substance along the lines of force; in this case the flux represents the amount of this substance displaced through each cross-section of the ring, and is analogous to total electrostatic displacement. Or, as some modern writers think, there is an actual flow of an incompressible ether along the lines of force. In that case the flux may be thought of as the rate of flow of the ether through a cross-section. The viewpoint common to these two explanations gave rise to the name flux which means flow.

Some physicists consider the magnetic circuit as consisting of infinitely subdivided (though closed) whirls or vortices in the ether, the rotation being in planes perpendicular to the lines of force. Each line of force is considered, then, as the geometric axis of an infinitely thin fiber or tube of force, and the ether within each tube in a state of transverse vortex motion. The line of force represents the direction of the axis of rotation, and the flux may be thought of as the momentum of the rotating substance per unit length of the tubes of force. According to any of these three views, the energy of a current is actually contained in the magnetic circuit linked with the current.

Whichever view is adopted, the magnetic flux can be defined as the sum total of magnetic disturbance through a cross-section perpendicular to the lines of force. Experiment shows that the total flux is the same through all complete cross-sections of a magnetic circuit. This could have been expected from the point of view of a displacement or flow along the lines of force; each tube of force being like a channel within which the displacement or the flow of an incompressible substance takes place. For this reason the magnetic flux is said to be *solenoidal* (i.e., channel-shaped).

The familiar law of electromagnetic induction discovered by Faraday is used for the definition of the unit of flux. Namely, when the total magnetic disturbance or flux within a turn of wire changes, an electromotive force is induced in the turn. By experiments in a uniform field, the fact is established that the value of the induced electromotive force is exactly proportional to the rate of change of the flux linking with the test loop. This fact is used in the definition of the unit of flux.

With the volt and the second as the units of e.m.f. and of time respectively, the corresponding unit of flux is called the *weber*, and is defined as follows: *A flux through a turn of wire changes at a uniform rate of one weber per second when the e.m.f. induced in the turn remains constant and equal to one volt.* Such a unit flux can be also properly called the *volt-second*, though as yet neither name has been recognized by the International Electrotechnical Commission. The weber or the volt-second is too large a unit for most practical purposes. Therefore a much smaller unit, called the *maxwell*,[1] is used, which is equal to one one-hundred-millionth part of the weber, or

$$\text{one maxwell} = \text{one weber} \times 10^{-8}.$$

The lines of force in Figs. 1 and 11 can be made to represent not only the direction of the field, but its magnitude as well, if they be drawn at suitable distances from each other. That is, such that the total number of lines passing through any part of a cross-section of the ring is equal numerically to the number of maxwells in the flux through the same part. With this convention, each line stands symbolically for one maxwell; some engineers and physicists speak of the number of lines of force in a flux when they mean maxwells.

While the weber is too large a unit, the maxwell is too small for many practical purposes. Therefore two other intermediate units

[1] The origin of the maxwell becomes clear when one remembers that the volt was originally established as 10^8 electromagnetic C.G.S. unit of electromotive force. The maxwell is related to the C.G.S. unit of e.m.f. or the so-called abvolt in the same way in which the weber is related to the ordinary volt. In other words, when the flux within a coil varies at the rate of one maxwell per second, one abvolt is induced in each turn of the winding.

are used, namely the kilo-maxwell, equal to one thousand max-wells, and the mega-maxwell, equal to one million maxwells. These two units are sometimes called the kilo-line and the mega-line, the word " line " being used for the word maxwell, as explained above.[1]

Prob. 1. The flux within the coil (Fig. 1) is equal to 63 kilo-maxwells. A test coil of five turns is wound on the exciting coil so as to be linked with the total flux. What voltage is induced in this test coil when the current in the main (exciting) coil is reduced to zero at a uniform rate in seven seconds?   Ans. 0.45 millivolt.

Prob. 2. At what rate must the flux be varied in the preceding problem in order to induce one volt in the test coil?
Ans. 0.2 weber (20 megalines) per second.

Prob. 3. When the flux varies at a non-uniform rate show that the voltage induced in the test coil at any instant is equal to $-n \, (d\Phi/dt)$ $\times 10^{-8}$, where n is the number of turns in series, t is time in seconds, and Φ is the flux in maxwells. Show that the exponent of 10 must be -2 instead of -8 if the flux is expressed in megalines.

4. The Reluctance of a Magnetic Path. Experiment shows that the total flux within the coil (Fig. 1) is proportional to the applied magnetomotive force, when the space inside is filled with air. Therefore, a relation similar to Ohm's law holds, namely,

$$M = \mathcal{R}\Phi, \quad \ldots \quad \ldots \quad \ldots \quad (1)$$

where M is the magnetomotive force in ampere-turns, Φ is the flux in maxwells, and $\mathcal{R}$ is the coefficient of proportionality between the two, called the *reluctance* of the magnetic circuit. Script $\mathcal{R}$ is used to distinguish reluctance from electric resistance. The mag-netomotive force M is the cause of the flux; or, with reference to an electric circuit, M is analogous to the applied electromotive force. Φ is analogous to the resulting current, and the reluctance $\mathcal{R}$ takes place of the electric resistance. Therefore, eq. (1) is known as Ohm's law for the magnetic circuit. Of course, the

---

[1] This possibility of creating new units of convenient size is a great advantage of the metric or decimal system of units. New units are gener-ally understood, by the use of Latin and Greek prefixes, signifying their numerical relation to the fundamental unit. For instance, it is perfectly legitimate to use such units as deci-ampere and hecto-volt, in spite of the fact that they are not in general use. Anyone familiar with the agreed prefixes will know that the units spoken of are equal to one-tenth of one ampere, and to one hundred volts. See Appendix I on the Ampere-Ohm System.

analogy is purely formal, the two sets of phenomena being entirely different. An equation similar to (1) can be written for the flow of heat, of water, etc. It merely expresses the experimental fact that, for a certain class of phenomena, the effect is proportional to the cause.

If the space within the coil be filled with practically any known substance, solid, liquid, or gaseous, the reluctance $\mathcal{R}$ remains within less than ± 1 per cent of the value which obtains with air. The notable exceptions are iron, cobalt, nickel, manganese, chromium, and some of their oxides and alloys.[1] When the circuit includes one of these so-called " ferro-magnetic " substances, a much larger flux is produced with the same m.m.f., that is, the reluctance of the circuit is apparently reduced to a considerable extent. Moreover, this reluctance is no longer constant, but depends upon the value of the flux. The behavior of iron and steel in a magnetic circuit is of great practical importance, and is treated in detail in Chapters II and III.

The definition of the unit of reluctance follows directly from eq. (1). A magnetic circuit has a unit reluctance when a magneto-motive force of one ampere-turn produces in it a flux of one maxwell.[2] No name has been given to this unit so far. The author ventures to suggest the name *rel*, and he uses it provisionally in this book. Granting that reluctance is a useful quantity in magnetic calculations, one must admit that it should be measured in some units of its own; unless one chooses to use the cumbersome notation " ampere-turns per maxwell." The name rel is simply the beginning of the word reluctance. Thus, a magnetic circuit has a reluctance of one rel when one ampere-turn produces one maxwell of flux in it. The unit rel is analogous to the ohm in the electric circuit, and to the daraf in the electrostatic circuit.

Prob. 4. What is the reluctance of the magnetic circuit in Fig. 1 if 47,600 ampere-turns produce a flux of 2.3 kilo-maxwells?
Ans. $47,600/2300 = 20.7$ rels.

Prob. 5. How many ampere-turns are required to establish a flux of 1.7 megalines through a reluctance of 0.0054 rel?   Ans.   9180.

Prob. 6. A wooden ring is temporarily wound with 330 turns of wire; when a current of 25 amperes is flowing through the winding the

[1] See Dr. C.P. Steinmetz, Magnetic Properties of Materials, *Electrical World*, Vol. 55 (1910), p. 1209.

[2] See Appendix I at the end of the book.

flux is found to be equal to 21 kilo-maxwells. The permanent winding on the same ring must produce a flux of 65.1 kilolines at a current of 9.3 amperes. How many turns will be required?   Ans. 2750.

5. The Permeance of a Magnetic Path. In calculations pertaining to the electric circuit it is convenient to deal with the reciprocals of resistances when conductors are connected in parallel. The reciprocal of a resistance is called a conductance and is measured in mhos if resistance is measured in ohms. Similarly, a dielectric is characterized sometimes by its elastance, at other times by the reciprocal of its elastance, which is called permittance. When permittance is measured in farads, elastance is measured in darafs (for the use of these units see Chapter XIV in the author's *Electric Circuit*).

Analogously, when two or more magnetic paths are in parallel it is convenient to use the reciprocals of the reluctances. The reciprocal of the reluctance of a magnetic path is called its *permeance*; eq. (1) becomes then

$$\Phi = \mathcal{P}M, \quad . \quad . \quad . \quad . \quad . \quad . \quad (2)$$

where

$$\mathcal{P} = 1/\mathcal{R}. \quad . \quad . \quad . \quad . \quad . \quad . \quad (3)$$

A script $\mathcal{P}$ is used for permeance in order to avoid confusing it with power. For the unit of permeance corresponding to the rel, the author proposes the name *perm*. A magnetic path has a permeance of one perm when one maxwell of flux is produced for each ampere-turn of magnetomotive force applied along the path.

The unit " perm " has been in use among electrical designers for some time, although no name has been given to it. Notably Mr. H. M. Hobart has used it extensively in his writings, in the calculation of the inductance of windings. He speaks of " magnetic lines per ampere-turn per unit length " (of the embedded part of a coil). This is equivalent to perms per unit length.

In the ampere-ohm system the internationally accepted unit of permeance is the *henry*.[1] Therefore, if in eq. (2) M is measured in ampere-turns and Φ in webers, $\mathcal{P}$ is in henrys, and no new unit for permeance is necessary. In this case the reluctance $\mathcal{R}$ in eqs.

---

[1] Although the henry is defined as the unit of inductance, it is shown in Art. 58 below that permeance and inductance are physically of the same dimensions and hence measureable in the same units.

(1) and (3) is in henrys^{-1}; or spelling the word henry backwards, as in the case of mho and daraf, the natural unit of reluctance in the ampere-ohm system gets the euphonious name of *yrneh* (to be pronounced earney).

Since, however, the maxwell is used almost exclusively as the unit of flux, it seems advisable to introduce the rel and the perm as units directly related to it. Should engineers gradually feel inclined to use the weber and its submultiples as the units of flux, then the henry, the yrneh, and their multiples and submultiples would naturally be used as the corresponding units of permeance and reluctance.

We have, therefore, the two following systems of units for reluctance and permeance, according to whether the maxwell or the weber is used for the unit of flux:

Unit of m.m.f.	Unit of flux	Unit of permeance	Unit of reluctance
Ampere-turn	Maxwell	Perm	Rel
Ampere-turn	Weber	Henry	Yrneh

One perm $= 10^{-8}$ henry; one rel $= 10^8$ yrnehs.

Prob. 7. What is the permeance of the magnetic circuit in prob. 4?
Ans. 0.0483 perm. $= 4.83 \times 10^{-10}$ henry.
Prob. 8. What is the permeance of the ring in prob. 6?
Ans. 2.545 perm. $= 0.02545$ microhenry.
Prob. 9. How many ampere-turns are required to maintain a flux of 2.7 megalines through a permeance of 750 perms?   Ans. 3600.

6. Reluctivity and Permeability. The reluctance of a magnetic path varies with the dimensions of the path according to the same law as the resistance of an electric conductor or the elastance of a dielectric. That is to say, the reluctance is directly proportional to the average length of the lines of force and is inversely proportional to the cross-section of the path. This relationship can be verified by measurements on rings of different dimensions (Fig. 1). When the diameter of the ring is increased twice, keeping the same cross-section, the length of the path of the flux is also increased twice. Experiment shows that the new ring requires twice as many ampere-turns as the first one for the same flux; or, only one-half of the flux is produced with the same number of ampere-turns.   If the diameter of the ring is kept the same but the cross-section of the path is increased twice, the flux is doubled with the same magnetomotive force.   These and similar experiments show that the reluctance and the permeance of a uniform magnetic path obey the same law as the resistance and the conductance of a conductor, or the elastance and the permittance of a prismatic slab of a dielectric.

We can therefore put

$$\mathcal{R} = \nu l / A, \quad . \quad . \quad . \quad . \quad . \quad . \quad . \quad (4)$$

where l is the mean length of the path, A is its cross-section, and ν is a physical constant.   By analogy with resistivity and elastivity, ν is called the *reluctivity* of a magnetic medium.   If $\mathcal{R}$ is in rels, and the dimensions of the circuit are in centimeters, ν is in rels per centimeter cube.   In other words, the reluctivity of a magnetic medium is the reluctance of a unit cube of this medium when the lines of force are parallel to one of the edges.   For air and all other non-magnetic substances the experimental value of ν is 0.8 rel per centimeter cube,[1] or 0.313 rel per inch cube.

The expression for permeance corresponding to eq. (4) is

$$\mathcal{P} = \mu A / l, \quad . \quad . \quad . \quad . \quad . \quad . \quad . \quad (5)$$

where the coefficient μ is called the *permeability* of the magnetic medium.   It corresponds to the electric conductivity γ and the dielectric permittivity κ.   Since the permeance of a path is the reciprocal of its reluctance, the permeability of a medium is the reciprocal of its reluctivity, or

$$\mu = 1/\nu. \quad . \quad . \quad . \quad . \quad . \quad . \quad (6)$$

When the perm and the centimeter are used for the units of permeance and length, permeability is expressed in perms per centimeter cube.   For all non-magnetic materials $\mu = 1.25$ perms per centimeter cube (more accurately 1.257).   With the henry and the centimeter as units $\mu = 1.257 \times 10^{-8}$ henries per centimeter cube.   In the English system $\mu = 3.19$ perms per inch cube for non-

---

[1] More accurately 0.796 rel per centimeter cube.   As a rule, magnetic calculations are much less accurate than electrical calculations, because there is no "magnetic insulator" known, so that there is always some magnetic leakage present, which is difficult to take into consideration.   For this reason the value 0.8 is sufficiently accurate for most practical purposes.

magnetic materials.   For magnetic materials μ is considerably larger than for non-magnetic, and varies with the field strength. In calculations either reluctivity or permeability is used, according to the conditions of the case and the preference of the engineer.

The student has probably heard before that the permeability of air is assumed equal to unity.   The discrepancy between this commonly accepted value and the value 1.25 given above, is due to a different unit of magnetomotive force, called the gilbert, which is sometimes employed.   The author considers the gilbert to be of doubtful utility, for reasons stated in Appendix II; hence no use is made of it in this book.

Prob. 10.   Assuming the value of $\mu = 1.257$ to be given, check the value of $\mu = 3.19$ in the English system, and also the values of ν in the metric and the English systems, as given above.

Prob. 11.   In prob. 4 the reluctance of a ring was 20.7 rels.   If the cross-section of the ring is 120 sq. mm., what is the average diameter of the ring?   Ans.   9.9 cm.

Prob. 12.   How many ampere-turns are required to establish a flux of 47 kilolines in a ring of rectangular cross-section, made of non-magnetic material; the radial thickness of the ring is 8 cm., the axial width 11 cm. and the average radius 16 cm?   Ans.   About 43 kiloampere-turns.

Prob. 13.   How many ampere-turns would be required in the preceding problem for the same flux if the ring were made of iron, the *relative* permeability of which (with respect to air) is 500?

Ans.   86 ampere-turns.

7. Magnetic Intensity.   In order that the student may better appreciate the significance of the concept of magnetic intensity, it is advisable to refresh in his mind the corresponding quantity used in the electric circuit, *viz.*, the electric intensity.   Namely, in problems on the electric and the electrostatic circuit it is sometimes desirable to consider not only the total voltage, but also the voltage used up or balanced per unit length of the path along which the electricity flows or is displaced.   This quantity, the rate of change of voltage along the circuit, is known as the electric intensity, or the voltage gradient.   It is denoted by G (see the *Electric Circuit*), and is measured in volts per linear centimeter. When the voltage drop is uniform along a conductor or a dielectric, $G = E/l$, where E is the voltage between the ends of the part of the circuit under consideration, and l is the corresponding length. When the voltage drop is not uniform, G is different for different points along the path, and for each point $G = dE/dl$.

In a similar way, the magnetomotive force of a magnetic circuit is used up bit by bit in the consecutive parts of the circuit. One can speak not only of the total magnetomotive force of a closed circuit, but also of the magnetomotive force acting upon a certain part of the circuit, and of magnetomotive force per unit length of the lines of force. Thus, for instance, if 1000 ampere-turns is consumed in a uniform magnetic circuit 4 cm. long, the magnetomotive force per unit length of path is 250 ampere-turns.

The magnetomotive force per unit length of path is called the magnetic intensity at a point, or the m.m.f. gradient, and is denoted by H. Thus, if the circuit is uniform, the magnetic intensity at any point is

$$H = M/l, \quad \ldots \ldots \ldots \quad (7)$$

where M is the magnetomotive force acting upon the length l of the circuit. If the magnetic circuit is non-uniform, for instance, if the cross-section of the ring is different at different places, or if the permeability is different at some parts of the circuit due to the presence of iron, the m.m.f. gradient is different at different points, and at each point it is expressed by the equation

$$H = dM/dl, \quad \ldots \ldots \ldots \quad (8)$$

where dM is the m.m.f. necessary for establishing the flux in the length dl of the circuit. If M is in ampere-turns, and l is in centimeters, H is in ampere-turns per centimeter.

Eqs. (7) and (8) can be also written in the form

$$M = Hl, \quad \ldots \ldots \ldots \ldots \quad (9)$$

and

$$M = \int_{l_1}^{l_2} H dl. \quad \ldots \ldots \ldots \quad (10)$$

These formulæ, expressed in words, simply mean that the magnetomotive force acting upon a certain part of a magnetic circuit is the *line integral* of the magnetic intensity along the path, or the sum of the m.m.fs. used up in the elementary parts of the path. The relation between M and H will become clearer to the student in the various applications that follow.

Prob. 14. What is the magnetic intensity in prob. 12?
Ans. About 425 ampere-turns per cm. of path.

8. Flux Density. It is often of importance to consider the *flux density*, or the value of a flux per unit of cross-section perpendicular to the direction of the lines of force. Flux density is usually denoted by B, and is measured in maxwells (or its multiples) per square centimeter.[1] When the flux is distributed uniformly over the cross-section of a path, the flux density

$$B = \Phi / A, \quad \ldots \ldots \ldots \quad (11)$$

where A is the area of the cross-section of the path. If the flux is distributed non-uniformly, an infinitesimal flux $d\Phi$ passing through a cross-section dA must be considered. In the limit, the flux density at a point corresponding to dA is

$$B = d\Phi / dA. \quad \ldots \ldots \ldots \quad (12)$$

The areas A and dA are understood to be at all points normal to the direction of the field. Solving these two equations for the flux we find

$$\Phi = B \cdot A, \quad \ldots \ldots \ldots \quad (13)$$

or

$$\Phi = \int_0^A B \cdot dA, \quad \ldots \ldots \quad (14)$$

the integration being extended over the whole cross-section of the path. Expressed in words, these last two formulæ mean that the total flux passing through a surface is equal to the sum of the fluxes passing through the different parts of that surface.

Magnetic flux density is analogous to current density U, and to dielectric flux density D treated in the *Electric Circuit*. The student will find no difficulty in interpreting eqs. (11) to (14) from the point of view of the electric and electrostatic circuits.

The relation between B and H is obtained from eq. (1) in which the value of $\mathcal{R}$ is obtained from eq. (4). Namely, we have

$$M = \Phi \nu l / A,$$

or

$$M / l = (\Phi / A)\nu.$$

[1] Some writers express flux density in gausses, one gauss being equal to one maxwell per square centimeter. The unit kilogauss, equal to one kilo-maxwell per square centimeter, is also used. While the terms gauss and kilogauss are convenient abbreviations, no use is made of them in this book in order to keep the relation between a flux and the cross-section of its path explicitly before the student.

The last expression, according to eqs. (7) and (11), can be written simply as

$$H = B\nu, \quad\quad \dotfill \quad (15)$$

or, since $\nu = 1/\mu$,

$$B = \mu H. \quad\quad \dotfill \quad (16)$$

Eqs. (15) and (16) state Ohm's law for a unit magnetic path, for instance, a path one centimeter long and one square centimeter in cross-section. H is the magnetomotive force between the opposite faces of the cube, μ is the permeance of the cube, and B is the flux passing through it. The reader will remember similar equations $U = \gamma G$ and $D = \kappa G$ for the unit electrical conductor and the unit prism of a dielectric respectively.

Instead of beginning the theory of the magnetic circuit with eq. (1) and developing it into eq. (16), it is possible to begin it with eq. (16). Namely, the known magnetic phenomena show that at each point in the medium there is a magnetic intensity H which is the cause of the magnetic state, and that the effect is measured by the flux density B; μ is the physical constant which shows the proportionality between H and B. The magnetic circuit is then assumed to be built up of infinitesimal tubes of flux in series and in parallel, and finally eq. (1) is obtained.

Prob. 15. What is the flux density in prob. 12?
    Ans. 534 maxwells per square centimeter (534 gausses).

Prob. 16. How many ampere-turns per pole are required to establish a flux density of 7 kilolines per square centimeter in the air-gap of a machine, the clearance being 3 mm.? Solution: According to eq. (15) $H = 7000/1.25 = 5600$ ampere-turns per centimeter of length. Hence the required m.m.f. is $5600 \times 0.3 = 1680$ ampere-turns.

9. Reluctances and Permeances in Series and in Parallel. In practice, one has to deal mostly with magnetic circuits of irregular form, for instance, those of electric machines (Fig. 24) in which the flux is established partly in air and partly in iron, each of varying cross-section. The circuit consists in this case of several reluctances in series. One may say, for instance, that the total magnetomotive force required in this machine, per magnetic circuit, is 8000 ampere-turns, of which 6000 are used in the air-gap, 1500 in the field frame, and 500 in the armature. This is analogous to distinguishing between the total e.m.f. of an electric circuit, and the voltage drop in the various parts of the circuit.

In some cases two or more magnetic paths are in parallel, for instance, when there is magnetic leakage (see below). In most cases the engineer has to consider complicated magnetic circuits which consist partly of paths in series, partly of paths in parallel. Thus, in the same machine, the m.m.f. or the *difference of magnetic potential* between the pole-tips is 6500 ampere-turns. This m.m.f. maintains a useful flux of say 2.5 megalines through the armature, and say 0.5 megaline of leakage or stray flux between the pole-tips. Thus the total flux in the field frame is 3 megalines.

The fundamental law of the magnetic circuit, as expressed by eq. (1), is analogous to Ohm's law for the simple electric circuit. Therefore magnetic paths in series and in parallel are combined according to the same rule that electrical conductors are combined in series and in parallel. Namely, when two or more magnetic paths are in series, their reluctances are added; when two or more magnetic paths are in parallel their permeances are added. Or, for a series combination,

$$\mathcal{R}_{eq} = \Sigma \mathcal{R}, \quad . \quad . \quad . \quad . \quad . \quad . \quad . \quad (17)$$

and for a parallel combination

$$\mathcal{P}_{eq} = \Sigma \mathcal{P}. \quad . \quad . \quad . \quad . \quad . \quad . \quad (18)$$

It will be remembered that similar relations hold also for impedances and admittances in the alternating current circuit, and for elastances and permittances in the electrostatic circuit.

The proof of formulæ (17) and (18) is similar to that usually given for the combination of electric resistances in series and in parallel. Namely, when reluctances are in series the total magnetomotive force is equal to the sum of component m.m.f.s., or

$$M_{eq} = \Sigma M. \quad . \quad . \quad . \quad . \quad . \quad . \quad (17a)$$

Dividing both sides of this equation by the common flux Φ eq. (17) is obtained. When permeances are in parallel, the total flux is the sum of the component fluxes, or

$$\Phi_{eq} = \Sigma \Phi. \quad . \quad . \quad . \quad . \quad . \quad . \quad (18a)$$

Dividing both sides of this equation by the common M, eq. (18) is obtained.

One of the reasons for which calculations are as a rule more involved and less accurate in the magnetic than in the electric circuit is *that there is no magnetic insulation known*, and therefore the paths of the flux in a great majority of cases cannot be shaped and confined at will. The student will appreciate, therefore, the reason for selecting a toroidal ring as the simplest magnetic circuit. If the winding is distributed uniformly there is no tendency for magnetic leakage, except for a very small amount in and around each wire. With almost any other arrangement of a magnetic circuit there is a difference of magnetic potential, or an m.m.f. between various parts of the circuit, and part of the flux passes directly through the path of the least resistance, in parallel with the useful path. A familiar example of this is the magnetic leakage between the adjacent pole-tips of an electrical machine (Fig.29), or between the coils of a transformer (Fig. 50).

The conditions in a magnetic circuit are similar to those in an imperfectly insulated electric circuit, when it, together with its sources of e.m.f., is immersed in a conducting liquid. Part of the current finds its path through the liquid instead of through the conductors; the current is different in different parts of the circuit, and the calculations are much more involved and less accurate, because the paths of the current in an unlimited medium can be estimated only approximately.

In order to prevent or to minimize leakage the exciting ampere-turns should be distributed over the whole magnetic circuit, to each part in proportion to its reluctance. Then the m.m.f. is consumed where it is applied, and no free m.m.f. is left for leakage. Unfortunately, such an arrangement is impracticable in most cases, though it ought to be approached as nearly as possible (see Prob. 17 below).

If there were a magnetic insulator, that is, a substance or a combination the permeability of which was many times lower than that of the air, it would be a great boon to the electrical industry. It would then be possible to avoid magnetic leakage by insulating magnetic circuits as perfectly as electric circuits are insulated. The absence of leakage would allow a reduction in the size of the field frames and exciting coils of direct- and alternating-current machines. It would also permit us to improve the voltage regulation of generators and transformers, to raise the power factor of induction motors, and to increase their overload capacity; it would also eliminate sparking in commutating machines.

Prob. 17. A long iron rod, having a cross-section of 9.3 sq.cm., is bent into a circular ring so that the ends almost touch each other. The ring is wound with 500 turns of wire, the winding being concentrated around the gap to minimize the leakage. When a current of 2.5 amperes is sent through the winding a flux of 74.9 kilo-maxwells is established in the circuit. Assuming the reluctance of the iron to be negligible, calculate the clearance between the ends of the rod.
Ans. Between 1.9 and 2.0 mm.

Prob. 18. What is the length of the air-gap in the preceding problem if the estimated reluctance of the iron part of the circuit is 2 milli-rels?
Ans. 1.7 mm.

Prob. 19. A magnetic circuit consists of three parts, the reluctances of which are $\mathfrak{R}_1 = 0.004$ rel, $\mathfrak{R}_2 = 0.005$ rel, and $\mathfrak{R}_3 = 0.013$ rel. The paths $\mathfrak{R}_2$ and $\mathfrak{R}_3$ are in parallel with each other and are in series with $\mathfrak{R}_1$. What is the total permeance of the circuit? Ans. 131.4 perms.

Prob. 20. In the preceding problem let $\mathfrak{R}_1$ be the reluctance of the steel frame of an electric machine, $\mathfrak{R}_2$ be that of two air-gaps, and the armature, and $\mathfrak{R}_3$ the leakage reluctance between two poles. The ratio of the total flux in the frame to the useful flux through the armature is called the leakage factor of the machine. What is its value in this case? Ans. 1.38.

Prob. 21. Referring to the two preceding problems let the air-gap be reduced so as to reduce the leakage factor to 1.2. How many ampere-turns will be required to produce a useful flux of 2.1 megalines in the magnetic circuit under consideration? Ans. 15,540.

Prob. 22. Let the cross-section of the magnetic circuit in Fig. 1 be 8 by 11 cm., and the mean diameter of the ring 32 cm. An iron ring having a cross-section of 4 by 5 cm., and the same mean diameter of 32 cm., is placed centrally (coaxially) within the coil. How many ampere-turns are required to produce a total flux of 47 kilolines (counting that in the air as well as that in the iron), if the estimated relative permeability of the iron is 1400? Hint: Let the average flux density in the air be B_a, and that in the iron be B_i. We have two simultaneous equations: $20B_i + (88-20)B_a = 47$, and $B_i/B_a = 1400$. Ans. 134.

Prob. 23. What per cent of the total flux in the preceding problem is in the air? Ans. 0.24 per cent.

Prob. 24. Show that in a ring, such as is shown in Fig. 1, the flux density, strictly speaking, is not uniform, but varies inversely as the distance from the center. Solution: Take an elementary tube of flux of a radius x. The magnetic intensity at any point within the tube is $H = M/2\pi x$, and the flux density, according to eq. (16), $B = \mu M/2\pi x$.

Prob. 25. What is the true permeance of a circular ring of rectangular cross-section, the outside diameter of which is D_1, the inside diameter D_2, and the axial width h? Solution: The permeance of an infinitesimal tube of radius x is $d\mathcal{P} = \mu h dx/2\pi x$. The permeances of all the tubes are in parallel and should be added; hence, integrating the foregoing expression between the limits $\frac{1}{2}D_1$ and $\frac{1}{2}D_2$ we get: $\mathcal{P} = (\mu h/2\pi)\mathrm{Ln}(D_1/D_2)$.

Prob. 26. Show that, when the radial thickness b of a ring is small as compared to its mean diameter D, the exact expression for permeance, obtained in the preceding problem, differs but little from the approximate value, $\mu h b/\pi D$, used before. Solution: Using the expansion, $\frac{1}{2}\mathrm{Ln}[(1+x)/(1-x)] = x + \frac{1}{3}x^3 + \frac{1}{5}x^5 + \ldots$ and putting

$$D_1/D_2 = (D+b)/(D-b) = (1+b/D)/(1-b/D);$$

we get $\mathcal{P} = (\mu h b/\pi D)[1 + \frac{1}{3}(b/D)^2 + \frac{1}{5}(b/D)^4 + \ldots]$. When the ratio of b to D is small, all the terms within the brackets except the first one, can be neglected.

Prob. 27. Show that the answer to prob. 12 is 2.1 per cent high on account of the density being assumed there as uniform throughout the cross-section of the ring.

CHAPTER II

THE MAGNETIC CIRCUIT WITH IRON

10. The Difference between Iron and Non-Magnetic Materials.
Steel and iron differ in their magnetic properties from most other
known materials in the following respects:

(1) The permeability of steel and iron is several hundred and
even thousand times greater than that of non-magnetic materials.

(2) The permeability of steel and iron is not constant, but
decreases as the flux density increases.

(3) Changes in the magnetization of steel and iron are
accompanied by some sort of molecular friction (hysteresis) with
the result that the same magnetomotive force produces a different
flux when the exciting current is increasing than when it is decreasing (Fig. 7).

Besides iron, the four adjacent elements in the periodic system,
viz., cobalt, nickel, manganese, and chromium, are slightly magnetic. Some alloys and oxides of these metals show considerable
magnetic properties. Heusler succeeded in producing alloys of
manganese, aluminum, and copper which are strongly magnetic.
These alloys have not been used in practice so far.[1]

11. Magnetization Curves. The magnetic properties of the steel
and iron used in the construction of electrical machinery are shown
in Figs. 2 and 3. These curves are called *magnetization curves*, or
B—H curves; sometimes also the *saturation curves of iron*.
The flux density, in kilolines per square centimeter of cross-section, is plotted, in these curves, against the ampere-turns per
centimeter length of the magnetic circuit as abscissæ.

The student may conveniently think of these curves as represent-

---

[1] For the preparation and properties of Heusler's alloys see Guthe and
Austin, *Bulletins of Bureau of Standards*, Vol. 2 (1906), No. 2, p. 297; Dr. C.
P. Steinmetz, *Electrical World*, Vol. 55 (1910), p. 1209; Knowlton. *Physical
Review*, Vol. 32 (1911), p. 54.

ing the results of tests on samples of iron in ring form, as in Fig. 1.[1] The current in the exciting coil is adjusted to a certain value, and the corresponding value of the flux in the iron ring is determined by any of the known means, for instance, by a discharge through

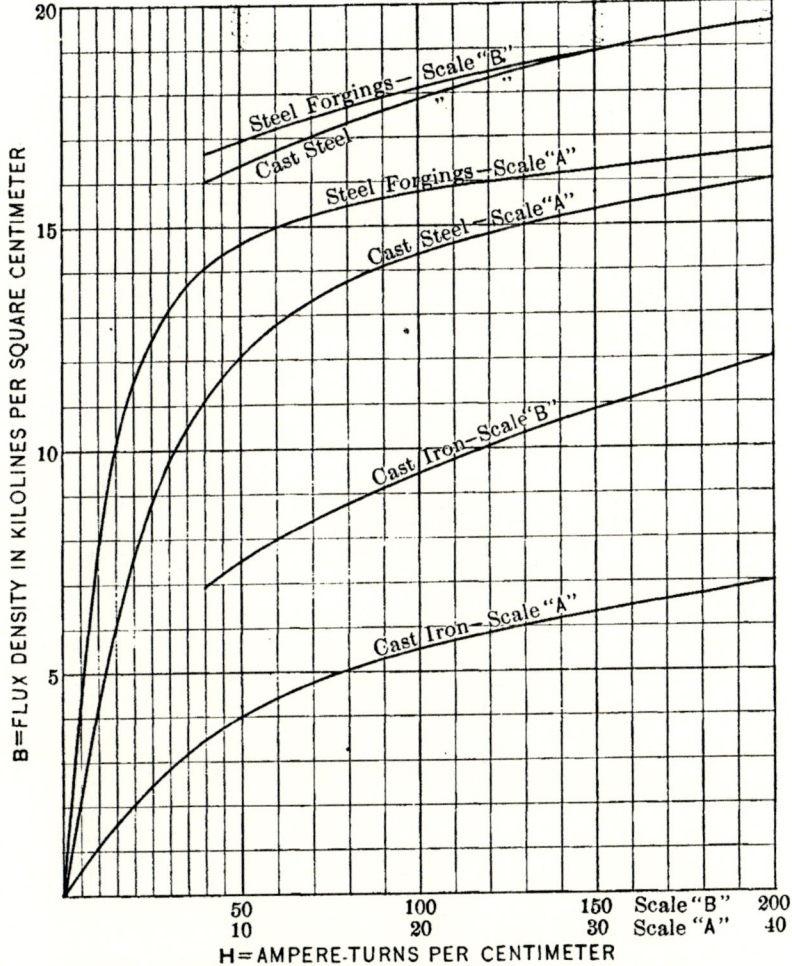

Fig. 2—Magnetization in steel and iron—castings and forgings.

[1] For an experimental study of the magnetic circuit with iron and for practical testing of the magnetic properties of steel and iron see Vol. 1, Chapters 6 and 7, of the author's *Experimental Electrical Engineering.*

a secondary coil connected to a calibrated ballistic galvanometer. The exciting ampere-turns divided by the average length of the path give the magnetic intensity H. The total flux divided by the cross-section of the iron path gives the value of the flux density B, which is plotted as an ordinate against H for an absissa. Similar tests are made for other values of H and B; the results give the magnetization curve of the material. In other words, a magnetization curve gives the relation between the magnetomotive force and the flux for a unit cube of the material. By combining unit cubes in series and in parallel a relationship is established between flux and ampere-turns for a circuit of any dimensions, made of the same material.

The curves shown in Fig. 2 refer to the following materials: (a) Cast iron, which is used as the magnetic material in the stationary field frames of direct-current machines, and in the revolving-field spiders of low speed alternators. It is evident from the curves that cast iron is magnetically much inferior to steel; but it is used on account of its lower cost and ease of machining. (b) Cast steel, which is used for pole pieces, plungers of electromagnets, etc. It is used also for the field frames of such machines in which economy of weight or space is desired, for instance, in railway and crane motors, and in machines built for export. (c) Forged steel, which is used for the revolving fields of turbo-alternators, on account of the considerable mechanical stresses developed in such high speed machines by the centrifugal force.

The curves in Fig. 3 refer to carbon-steel laminations and to silicon-steel laminations. The former is used in the armatures of direct and alternating-current machines, the latter mainly in transformers. There is not much difference between the two kinds with regards to their B—H curves, but silicon steel shows a much lower loss of energy due to hysteresis and eddy currents (see Art. 20 below). A material of much higher permeability is used for armature cores, when it is desired to use very high flux densities in the teeth. A magnetization curve for such steel laminations is shown in Fig. 28.

For convenience and accuracy the lower part of each curve in Fig. 2 is plotted separately to a larger scale, " A," while the upper parts are plotted to a smaller scale, " B." Thus, Fig. 2 contains only three complete magnetization curves. The curve for silicon-steel laminations in Fig. 3 is also plotted to two different scales,

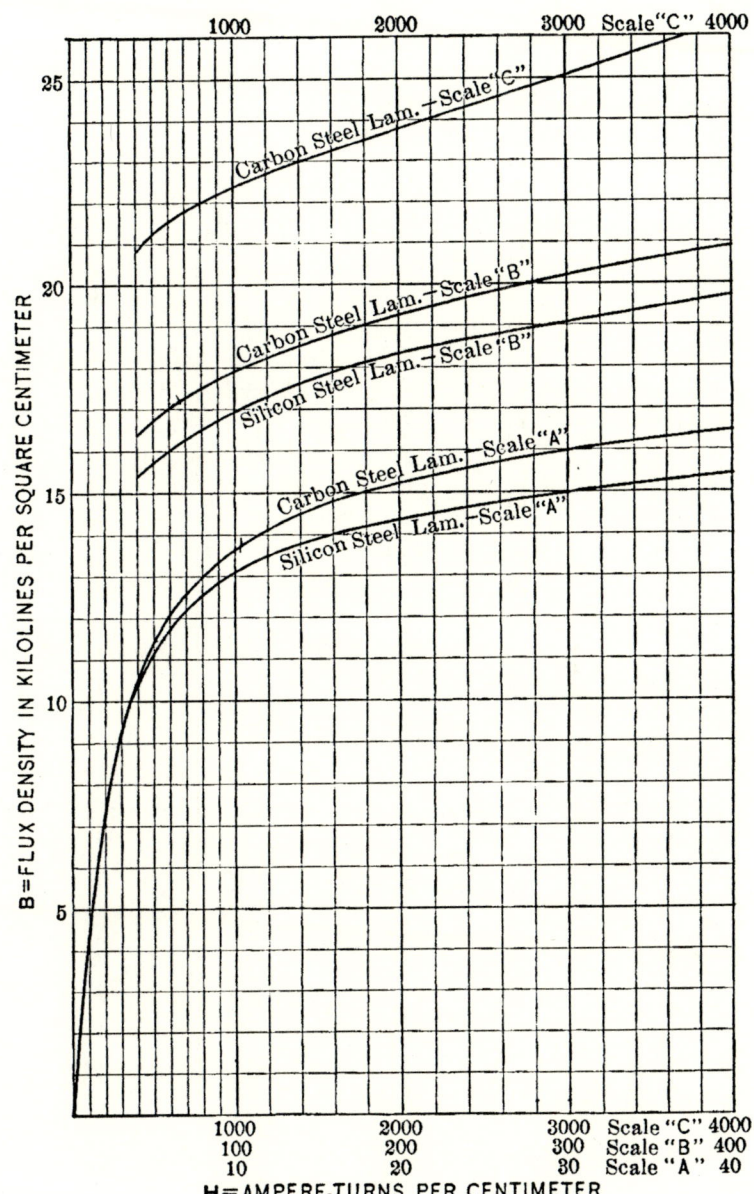

FIG. 3.—Magnetization in steel laminations.

while three different scales are used for the carbon-steel curve.
The values of H at very low flux densities are unreliable because
in reality each curve has a point of inflexion near the origin, not
shown in Figs. 2 and 3 (see Fig. 7).

The curves given in Figs. 2 and 3 represent the averages of many
curves obtained from various sources. The iron used in an individual case may differ considerably in its magnetic quality from the
average curve. The value of B obtainable with a given H depends
to a large degree upon the chemical constitution of the specimen,
impurities, heat treatment, etc. As a rule, the soft and pure grades
of steel are magnetically better, that is to say, they give a higher
flux density for the same magnetizing force, or, what is the same,
they possess a higher permeability. Annealing improves the
magnetic quality of iron, while punching, hammering, etc., lowers
it. Therefore, the laminations used in the construction of electrical machinery are usually annealed after being punched into
their final shape. This annealing also reduces hysteresis loss.

12. Permeability and Saturation. Permeability is defined in
Chapter I as the permeance of a unit cube, or, according to eq.
(16), as the ratio of B to H. The two definitions are, of course,
identical. Therefore, the values of permeability for various values
of B are easily obtained from the magnetization curves. For
instance, for cast steel, at $B = 15$ kilolines per sq. cm. the magnetic
intensity H is 26 ampere-turns per cm., so that $\mu = 15000/26 = 577$
perms per cm. cube. This is the value of the *absolute permeability in the ampere-ohm-maxwell system.* In most books the relative
permeability of iron is employed, referring to that of the air as
unity. Since in the above-mentioned system $\mu = 1.25$ for air,
the relative permeability of cast steel at the selected flux density
is $577/1.25 = 461$.

In practice, the calculations of magnetic circuits with iron are
arranged so as to avoid the use of permeability μ altogether, using
the B-H curves directly. In some special investigations, however, it is convenient to use the values of permeability, and also
an empirical equation between μ and B. For instance, see the
Standard Handbook for Electrical Engineers; the topic is indexed
" permeability—curves," and " permeability—equation." These
$\mu - B$ curves show that there must be a point of inflection in
the $B - H$ curves at low densities, because the values of μ reach
their maximum at a certain definite density instead of being con-

stant for the lower part of the curves. Such would be the case if the lower parts of the $B-H$ curves were straight lines, as shown in Figs. 2 and 3, because then the ratio of B to H would be constant. However, in ordinary engineering work the lower parts of magnetization curves are usually assumed to be straight lines, and the permeability constant.

Three parts can be distinguished in a $B-H$ or magnetization curve: the lower straight part, the middle part called the knee of the curve, and the upper part, which is nearly a straight line. As the magnetic intensity H increases, the corresponding flux density B increases more and more slowly, and the iron is said to approach *saturation*. With very high values of the magnetic intensity H, say several thousand ampere-turns per centimeter, the iron is completely saturated and the rate of increase of flux density with H is the same as in air or in any other non-magnetic material. That is to say, the flux density B increases at a rate of 1.257 kilolines for each kilo-ampere-turn increase in H. Such is the slope of the upper curve in Fig. 3.

In view of this phenomenon of saturation the total flux density in iron can be considered as consisting of two parts, one due to the presence of iron, the other independent of it, as if the paths of the lines of force were in air. These two parts are shown separately in Fig. 4. The part OA, due to the iron, approaches a limiting value B_s, where the iron is saturated. The part OC, not due to the iron, increases indefinitely in accordance with the straight line law, $B = \mu H$, where $\mu = 1.257$. The curve OD of total flux density resembles in shape that of OA, but approaches asymptotically a straight line KL parallel to OC.

While it is customary to speak of the saturation in iron as being low, high, or medium, the author is not aware of any generally recognized method of expressing the degree of saturation numerically. It seems reasonable to define per cent saturation in iron with respect to the flux density B_s, so that, for instance, the per cent saturation at the point N is equal to the ratio of PN' to B_s. This method of defining saturation, while correct theoretically, presupposes that the ordinate B_s is known, which is not always the case.

The percentage saturation of a machine is defined in Art. 58 of the *Standardization Rules* of the American Institute of Electrical Engineers (edition of 1910) as the percentage ratio of OQ to PN,

QN being a tangent to the saturation curve at the point N under consideration. An objection to this definition is that according to it the per cent saturation does not approach 100 as N increases indefinitely; on the contrary, the per cent saturation gradually decreases to zero beyond a certain value of N. This is, of course, absurd. Moreover, the foregoing definition of the Institute refers explicitly to the " percentage of saturation of a machine," and it is not clear whether magnetization curves of the separate materials are included in it or not. The practical advantage of this definition as compared to that given above is that it is not necessary to know the value of B_s.

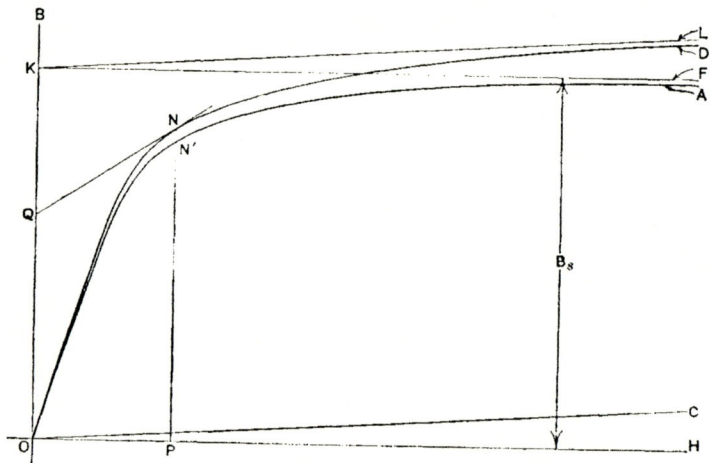

Fig. 4.—A magnetization curve analyzed.

13. Problems Involving the Use of Magnetization Curves. The following problems have been devised to give the reader a clear understanding of the meaning of magnetization curves, and to develop fluency in their use. These problems lead up to the magnetic circuit of electric machines treated in Chapters V and VI. With almost any arrangement of a magnetic circuit there is some leakage or spreading of the lines of force, which is difficult to take into account theoretically. This leakage is neglected in most of the problems that follow, so that the results are only approximately correct. Leakage is considered more in detail in Art. 40 below, though practical designers are usually satisfied with estimating it from the results of previous tests, rather than to calculate it theoretically.

Prob. 1. Samples of cast steel are to be tested for their magnetic quality up to a density of 19 kilolines per square centimeter. They are to be in the form of rings, 20 cm. average diameter, and 0.75 sq.cm. cross-section. For how many ampere-turns should the exciting winding be designed, and what is the lowest permeance of the circuit, if some specimens are expected to have a permeability 10 per cent lower than that according to the curve in Fig. 2?

Ans. 40.4 kiloampere-turns; 1.37 perm.

ν **Prob. 2.** Explain the reason for which it is not necessary to know the cross-section of the specimens in order to calculate the necessary ampere-turns in the preceding problem.

ι **Prob. 3.** Some silicon steel laminations are to be tested in the form of a rectangular bunch 20 by 2 by 1 cm., in an apparatus called a permeameter. The net cross-section of the iron is 90 per cent of that of the packet. It is found for a sample that 336 ampere-turns are required to produce a flux of 25.2 kilo-maxwells, the ampere-turns for the air-gaps and for the connecting yoke of the apparatus being eliminated. How does the quality of the specimen compare with the curve in Fig. 3?

Ans. The permeability of the sample at $B = 14$ is about 5 per cent lower than that according to the curve.

Prob. 4. What are the values of the absolute and the relative permeability and reluctivity of the sample in the preceding problem?

Ans. μ ν

relative 663 (numeric)   0.00151 (numeric)

absolute 833 perms per cm. cube   0.00120 rels per cm. cube.

Prob. 5. What is the maximum permeability of cast iron according to the curve in Fig. 2?   Ans.   About 600 perms per cm. cube.

Prob. 6. Mark in Figs. 2 and 3 vertical scales of absolute and relative permeability, so that values of permeability could be read off directly by laying a straight edge between the origin and the desired point of the magnetization curve.

ν **Prob. 7.** What is the percentage of saturation in carbon steel laminations at a flux density of 20 kilo-maxwells per square centimeter, according to both definitions given in Art. 12?   Ans. 92.5; 88.

ι **Prob. 8.** An electromagnet has the dimensions (in cm.) shown in Fig. 5; the core is made of carbon steel laminations 4 mm. thick, the lower yoke is of cast iron. The length of each air-gap is 2 mm.; each exciting coil has 450 turns. What is the exciting current for a useful flux of 2.2 megalines in the lower yoke? Neglect the magnetic leakage between the limbs of the electromagnet (this leakage is taken into consideration in the next problem). Solution: With laminations 4 mm. thick the space occupied by insulation between stampings is altogether negligible; therefore the flux density in the steel is the same as in the air gap and is equal to 17.2 kl/sq. cm.; in the cast iron the flux density is 11.5

kl/sq. cm. One-half of the average length of the path in the steel is 37.3 cm., and in the cast iron 20.5 cm. Hence, with reference to the magnetization curves, we find for one-half of the magnetic circuit (the other half being identical, it is sufficient to calculate for one-half):

amp.-turns for steel core $65 \times 37.3 = 2425$
amp.-turns for one air-gap $0.2 \times 0.8 \times 17200 = 2752$
amp.-turns for the cast-iron yoke $180 \times 20.5 = 3690$

Total   8867.

Ans.   The exciting current is 8867/450 = 19.7 amperes.

Prob. 9. In the solution of the preceding problem the effect of leakage is disregarded. It is found by experiments on similar electromagnets

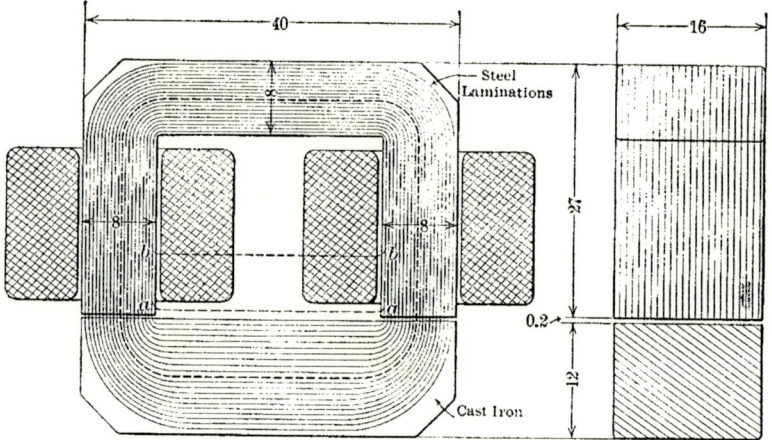

Fig. 5.—An electromagnet (dimensions in centimeters).

that the leakage factor is equal to about 1.2, that is to say, the flux in the upper yoke is 20 per cent higher than that in the lower one. This means that out of every 1200 lines of force in the upper yoke 1000 pass through the lower yoke as a part of the useful flux, and 200 find their path as a leakage through the air between the limbs, as shown by the dotted lines. Calculate the exciting current required in the preceding problem, assuming (a) that the total leakage flux is concentrated between the two air-gaps along the line *aa*; (b) that it is concentrated along the line *bb*, at one-third of the distance from the bottom of the exciting coil, that is 6.33 cm. from the air-gaps.

Ans.   (a) 44.2 amperes; (b) 40 amperes.

Prob. 10. Show that it is more correct in the preceding problem to assume the leakage flux concentrated at one-third of the distance from the bottom of the exciting coils, than at the center of the coils.

Prob. 11. A ring of forged steel has such dimensions that the average length of the lines of force is 70 cm. The ring has an air-gap of 1.5 mm., and is provided with an exciting winding concentrated near the air-gap so as to minimize the leakage. What is the flux density at an m.m.f. of 4000 ampere-turns? First Solution: Assume various values of B, calculate the corresponding values of the ampere-turns, until the value of B is found, for which the required excitation is 4000 ampere-turns (solution by trials). Second solution: Let the unknown density be B and the corresponding magnetic intensity in the steel be H. The required excitation for the steel is then $70H$, and for the air-gap $0.15 \times 0.8 \times 1000B = 120B$ ampere-turns. Therefore,

$$70H + 120B = 4000.$$

The values of B and H must satisfy this equation of a straight line, and besides they must be related to each other by the magnetization curve for steel forgings (Fig. 2). Hence, B and H are determined by the intersection of the straight line and the curve. The straight line is determined by two of its points; for instance, when $H = 40$, $B = 10$; when $H = 24$, $B = 19.3$. Drawing this line in Fig. 2 we find that the point of intersection corresponds to $B = 16.3$.[1]   Ans.   16.3 kilolines per sq. cm.

Prob. 12. Solve the preceding problem, assuming the ring to be made of silicon steel laminations: 10 per cent of the space is taken by the insulation between the laminations.

Ans.   Flux density in the laminations is 15.2 kl/sq. cm.

Prob. 13. In a complex magnetic circuit, an air-gap 3 mm. long and 26 sq. cm. in cross-section is shunted by a cast-iron rod 14 cm. long and 10 sq. cm. in cross-section. What is the number of ampere-turns necessary for producing a total flux of 215 kilolines through the two paths in parallel, and what is the reluctance of the rod per centimeter of its length under these conditions?   Ans.   1160 ampere-turns; 0.933 milli-rel.

Prob. 14. The magnetic flux in a closed iron core must increase and decrease according to a straight-line law with the time, then reverse and increase and decrease according to the same law in the opposite direction. Show the general shape of the curve of the exciting current, neglecting the effect of hyteresis.

Prob. 15. Show that if in the preceding problem the flux varies according to the sine law the curve of the exciting current is a peaked wave. Show how to determine the shape of this curve from a given magnetization curve of the material. This problem has an application in the calculation of the exciting current in a transformer.

Prob. 16. In the magnetic circuit shown in Fig. 6 the useful flux passes through the air-gap between the two steel poles; a part of the flux

---

[1] The student will see from the solution of this problem that in the case of a series magnetic circuit it is much easier to find the m.m.f. required for a given flux than *vice versa*. On the other hand, in the case of two magnetic paths in parallel (such as in prob. 13), it is easier to find the flux for a given m.m.f.

is shunted through the cast-iron part of the circuit. At low saturations a considerable part of the total flux is shunted through the cast-iron part, but as the flux density increases the cast iron becomes saturated, and a larger and larger portion of the flux is deflected into the air-gap. What percentages of the total flux in the yoke are shunted through the cast iron when the flux density in the air-gap is 1 kl/sq. cm. and 7 kl/sq. cm. respectively? Solution: When the flux density in the air-gap is 1 kilo-line per sq. cm. the m.m.f. across the gap is $1000 \times 0.8 \times 0.5 = 400$ ampere-turns. The flux density in the steel poles is 2 kl/sq. cm., and the required m.m.f. in them is about 16 ampere-turns. Therefore, the total m.m.f. across AC and consequently across the cast-iron part is 416 ampere-turns,

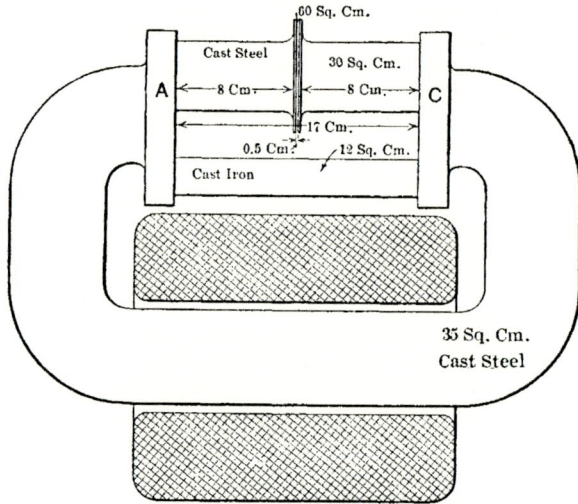

FIG. 6.—A complex magnetic circuit.

or $H = 24.5$ ampere-turns per centimeter of length of the path in the cast iron. This value of H corresponds on the magnetization curve to $B = 6$ kl/sq. cm.; hence, the total flux in the cast iron is 72 kl. The flux in the yoke is $60 + 72 = 132$ kl., and the percentage in the cast-iron shunt is $72/132$ or about 55 per cent. Similarly, it is found that, when the flux density in the air-gap is 7/kl sq. cm., about 25 per cent of the flux is shunted through the cast-iron part. The foregoing arrangement illustrates the principle used in some practical cases, when it is desired to modify the relation between the flux and the magnetomotive force, by providing a highly saturated magnetic path in parallel with a feebly saturated one.

Ans. 55 per cent and 25 per cent approximately.

Prob. 17. Indicate how the preceding problem can be solved if the cast-iron part were provided with a small clearance of say 1 mm. Hint: See the second solution to problem 11.

✓Prob. 18. What is the length of the yoke in Fig. 6 if the exciting current increases 12 times when the flux density in the air-gap increases from 1 to 7 kl/sq. cm.?   Hint: If H_1 and H_7 are the known magnetic intensities in the yoke, corresponding to the two given densities, and x is the unknown length of the yoke, we have, using the values obtained in the solution of problem 15: $(H_1 x + 416)12 = H_7 x + 3090$.

Ans.   About 1.2 m.

CHAPTER III

HYSTERESIS AND EDDY CURRENTS IN IRON

14. The Hysteresis Loop. Steel and iron possess a property of retaining part of their magnetism after the external magnetomotive force which magnetized them has been removed. Therefore, the magnetization or the B–H curve of a sample depends somewhat upon the magnetic state of the specimen before the test. This property of iron is called *hysteresis*. The curves shown in Figs. 2 and 3 refer to the so-called *virgin* state of the materials, which state is obtained by thoroughly demagnetizing the sample before the test. A piece of iron can be reduced to the virgin state by placing it within a coil through which an alternating current is sent, and gradually reducing the current to zero. Instead of changing the current, the sample can be removed from the coil.

Let a sample of steel or iron to be tested be made into a ring and provided with an exciting winding, as in Fig. 1. Let it be thoroughly demagnetized; in other words, let its *residual* magnetism be removed; then let the ring be magnetized gradually or in steps to a certain value of the flux density. Let OA in Fig. 7 represent the virgin magnetization curve, that is to say the relation between the calculated values of B and H from this test, and let PA be the highest flux density obtained. If now the magnetizing current be gradually reduced, the relation between B and H is no more represented by the curve OA, but by another curve, such as AC; this is because of the above-mentioned property of iron to retain part of its magnetism. When the current is reduced to zero, the specimen still possesses a *residual* flux density OC. Let the current now be reversed and increased in the opposite direction, until H reaches the negative value OF, at which no magnetic flux is left in the sample. The value of $H = OF$ is called the *coercive force*. When the magnetic intensity reaches the negative value of $O\overset{\circ}{P}{}' = OP$, experiment shows that the magnetic density $P'A'$ in the sample is equal and opposite to PA.

32

Let now the exciting current be again decreased, reversed and increased to its former maximum value corresponding to $H = OP$. It will be found that the relation between B and H follows a different though symmetrical curve, $A'C'F'A$, which connects with the upper curve at the point A. The complete closed curve is called the *hysteresis loop*; a sample of iron which has been subjected to a varying magnetomotive force as described before, is said to have undergone a complete cycle of magnetization. If the same cycle

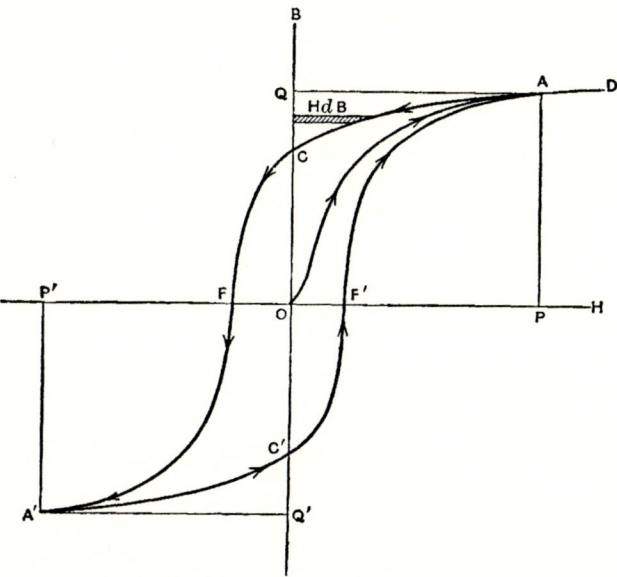

FIG. 7.—A hysteresis loop.

be repeated any number of times, the curve between B and H remains the same, as long as the physical properties of the sample remain unchanged.

The lower half of the hysteresis loop is identical with the inverted upper half, so that the residual flux density $OC' = OC$, and the coercive force $OF' = OF$. The shape of the loop for a given sample is completely determined by the maximum ordinate AP, or the maximum excitation OP. If the excitation be carried further, for instance, to the point D on the virgin curve, the hysteresis loop would be larger, beginning at the point D, and would be similar in its general shape to the loop shown in Fig. 7.

A piece of iron can also be carried through a hysteresis cycle mechanically. Thus, instead of changing the excitation, the sample may be moved to a weak field, reversed, and returned to its original location. The relation between B and H, however, will be the same in either case.

An important feature of the hysteresis cycle is that it requires a certain amount of energy to be supplied by the magnetizing current, or by the mechanism which reverses the iron with respect to the field. It is proved in Art. 16 below that this energy per cubic unit of iron is proportional to the area of the hysteresis loop. This energy is converted into heat in the iron, and therefore from the point of view of the electromagnetic circuit represents a pure loss. If the cycles of magnetization are performed in sufficiently rapid succession, for instance by using alternating current in the exciting winding, the temperature of the iron rises appreciably.

The phenomenon of hysteresis is irreversible; that is to say, it is impossible to make a piece of iron to undergo a cycle of magnetization in the direction opposite to that indicated by arrowheads in Fig. 7. If it were reversible, the loss of energy occasioned by performing the cycle in one direction could be regained by performing it in the opposite direction. In this respect the hysteresis cycle differs materially from the theoretical reversible cycles studied in thermodynamics, and reminds one of an irreversible thermodynamic cycle, in which friction or sudden expansion is present.

15. An Explanation of Saturation and Hysteresis in Iron. While the physical nature of magnetism is at present unknown, there is sufficient evidence that the magnetization of iron is accompanied by some kind of molecular change. Let us assume, in accordance with the modern electronic theory, that there is an electric current circulating within each molecule of iron, due to the orbital motion of one or more electrons within the molecule. Each molecule represents, therefore, a minute electromagnet acted upon by other molecular electromagnets. In the neutral state of a piece of iron, the grouping of the molecules is such that the currents are distributed in all possible planes, and the external magnetic action is zero. Under the influence of an external magnetomotive force the molecules are oriented in the same way that small magnetic needles are deflected by an external magnetic field. With small intensities of the external field, the molecules of iron return into their original stable positions as soon as the external m.m.f. is removed; when, however, the external magnetic intensity becomes considerable some of the molecules turn violently and assume new groupings of stable equilibrium. Therefore, when the external m.m.f. is removed, there is some intrinsic magnetization left, and we have the phenomenon of residual magnetism.

With an ever-increasing external m.m.f., more and more of the molecules are oriented so that their m.m.fs. are in the same direction as the external field, the iron then approaching saturation. Any further increase in the flux density is then mainly due to the flux between the molecules, the same as in any non-magnetic medium.

According to the foregoing theory, an external m.m.f. turns the internal m.m.fs. into more or less the same direction; these m.m.fs. then help to establish the flux in the intermolecular spaces which are much greater than the molecules themselves. Therefore, the higher flux density in iron is not due to a greater permeability of the iron itself, but to an increased m.m.f. It is nevertheless permissible, for practical purposes, to speak of a higher permeability of the iron, disregarding the internal m.m.fs., and considering the permeability, according to eq. (16), as the ratio of the flux density to the *externally* applied magnetic intensity.

The foregoing theory explains also the general character of the permeability curve of iron. With very small values of H the molecules of a piece of iron are oriented but very little, but are rapidly oriented more and more as H is increased. Therefore, for small values of H, μ must be expected to increase with H. On the other hand, when the saturation is very high, an increase in H changes B but little, because practically all of the available internal m.m.fs. have been utilized. Therefore, for large values of H, μ decreases with increasing H. Consequently, there is a value of H for which μ is a maximum. This is the actual shape of permeability curves (see for instance the reference to the *Standard Handbook* given in Art. 12 above).

The phenomenon of magnetization is irreversible because the changes from one stable grouping of molecules to the next are sudden. Each molecule, in changing to a new grouping, acquires kinetic energy, and oscillates about its new position of equilibrium until the energy is dissipated by being converted into heat. This heat represents the loss of energy due to hysteresis.

This theory of saturation and hysteresis is due originally to Weber, and has been improved by Ewing, who has shown experimentally the possibility of various stable groupings of a large number of small magnets in a magnetic field. By varying the applied m.m.f. he obtained a curve similar to the hysteresis loop of a sample of iron. For further details of this theory see Ewing, *Magnetic Induction in Iron and other Metals* (1892), Chapter XI.

The following analogy is also useful. Let a body Q (Fig. 8), rest on a support and be held in its central position by two springs

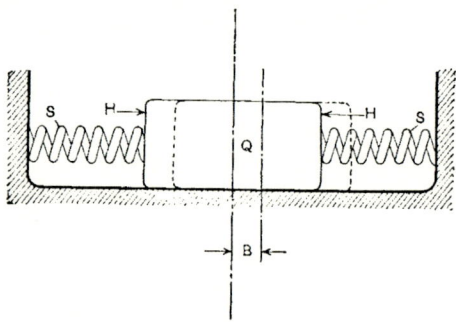

FIG. 8.—A mechanical analogue to hysteresis.

S, S, which can work both under tension and under compression. Let this body be made to move periodically to the right and to the left of its central position, under the influence of an alternating external force H. Call B the deflections of the body from its middle position. The relation between B and H is then similar to the hysteresis loop in Fig. 7, provided that there is some friction between the body Q and its support, and provided that the springs offer in proportion more resistance when distorted greatly than when distorted slightly.

Starting with the neutral position of the body let a gradually increasing force H be applied which moves the body to the right. This corresponds to the virgin curve in Fig. 7, except that this simple analogy does not account for the inflection in the virgin curve near the origin. Let then the force H be gradually reduced, allowing the springs to bring Q nearer the center. When the external force is entirely removed, the body is still somewhat to the right of its central position, because the friction balances part of the tension of the springs. Here we have something analogous to residual magnetism and to the part AC of the hysteresis loop. A finite force H is required in the negative direction to bring Q to the center. This force corresponds to the coercive force of a piece of iron.

By following this analogy through the complete cycle one can show that a loop is obtained similar to a hysteresis loop. Also, it can be shown that the phenomenon is irreversible, and that the total work done by the force H is equal to the work of friction. Moreover, there is a periodic interchange of energy between the springs and the source of the force H, and the net loss of energy is represented by the area of the loop corresponding to Fig. 7.

Prob. 1. An iron ring is thoroughly demagnetized, and then the current in the exciting winding is varied in the following manner: It is increased gradually from zero to 1 ampere and is then reduced to zero. After this, the current is increased to 2 amperes in the same direction, and again reduced to zero. Then the current is increased to 3 amperes again in the same direction, and reduced to zero, etc. Draw roughly the general character of the B-H curve, taking the hysteresis into consideration. Hint: First study a similar process on the mechanical analogy shown in Fig. 8.[1]

Prob. 2. A piece of iron is made to undergo a magnetization process from the point A (Fig. 7) to a point between F and A' such that, when subsequently the exciting circuit is opened, the ascending branch of the hysteresis curve comes to the origin. Show that such a process does not bring the iron into the neutral virgin state, in spite of the fact that $B = 0$ for $H = 0$. Hint: Consider the further behavior of the iron for positive and negative values of H.

Prob. 3. A millivoltmeter is connected to the high-tension terminals of a transformer, and the current in the low-tension winding is varied in such a way as to keep the voltage constant. Show that the curve of the current plotted against time is of the same shape as the hysteresis loop of the core. Hint: Since $d\Phi/dt$ is constant, Φ is proportional to the time.

Prob. 4. The magnetic flux density in an iron core is to vary with the time according to the sine law. Plot to time as abscissæ the instantaneous values of the exciting ampere-turns per centimeter length of the core from an available hysteresis loop, and show that the wave of the exciting current is not a sine wave and is unsymmetrical. Note: This problem has an application in the calculation of the exciting current of a transformer; see Art. 33 below.

[1] A solution of this and of the next problem will be found in Chapter V of Ewing's *Magnetic Induction in Iron and other Metals*, 1892.

16. The Loss of Energy per Cycle of Magnetization. When a magnetic flux is maintained constant the only energy supplied from the source of electric power is that converted into the i^2r heat in the exciting winding; no energy is necessary to *maintain* the magnetic flux. This is an experimental fact, fundamental in the theory of magnetic phenomena. When, however, the flux is made to vary, by varying the exciting ampere-turns or the reluctance of the magnetic circuit, electromotive forces are induced in the magnetizing, winding by the changing flux. A transfer of energy results between the electric and the magnetic circuits.

Beginning, for instance, at the point A of the cycle (Fig. 7), and going toward C, the flux is forced to decrease. According to Faraday's law, the e.m.f. induced by this flux in the magnetizing winding is such as to resist the change, i.e., it tends to maintain the current. Therefore, during the part AC of the cycle energy is supplied from the magnetic to the electric circuit. This shows that energy is stored in a magnetic field. During the part CFA' of the hysteresis loop energy is supplied from the electric to the magnetic circuit, because at the point C, the current is reversed and becomes opposed to the e.m.f. The other half of the cycle being symmetrical, with the flux and the current reversed, energy is returned to the electric circuit during the part $A'C''$ of the cycle, and is again accumulated in the magnetic circuit during the part $C'F'A$.

If the part AC of the cycle were identical with $C'F'A$, and the part $A'C''$ were identical with CFA', the amounts of energy transferred both ways would be the same, and there would be no net loss of energy at the end of the cycle. In reality the two parts are different; the amounts of energy returned from the magnetic circuit to the electric circuit in the parts AC and $A'C''$ are smaller than the amounts supplied by the electric circuit in the parts CFA' and $C'F'A$. This is because the last two parts of the curve are more steep than the first two, and consequently the induced e.m.fs. are larger for the same values of the current. The net result is therefore an input of energy from the electric into the magnetic circuit, this energy being converted into heat in the iron. No such effect is observed with non-magnetic materials, because the two branches of a complete B-H cycle coincide with a straight line passing through the origin.

To prove that the energy lost per cubic unit of iron per cycle of magnetization is represented by the area of the hysteresis loop, we first write down the expression for the energy returned to the electric circuit during an infinitesimal change of flux in the part AC of the cycle. Let the flux in the ring at the instant under consideration be Φ webers, and the magnetomotive force ni ampere-turns, where i is the instantaneous value of the current, and n is the total number of turns on the exciting winding. The instantaneous induced e.m.f., due to a decrease of the flux by $d\Phi$ during an infinitesimal element of time dt seconds, is $e = -nd\Phi/dt$ volt. The sign minus is necessary because e is positive (in the direction of the current) when $d\Phi$ is negative, that is to say, when the flux decreases.   The electric energy corresponding to this voltage is

$$dW = eidt = -nid\Phi \text{ watt-seconds (joules)}.$$

Hence, the total energy returned to the electric circuit during the part AC of the cycle is

$$W = -\int_{A}^{C} nid\Phi,$$

or, interchanging the limits of integration,

$$W = \int_{C}^{A} nid\Phi.$$

Since all the parts of the ring undergo the same process, and the curve in Fig. 7 is plotted for a unit cube of the material, it is of interest to find the loss of energy per cubic centimeter of material.   If S is the cross-section and l the mean length of the lines of force in the iron, we have that the volume

$$V = Sl \text{ cubic centimeters.}$$

Dividing the expression for the energy by this equation, we find that the energy in watt-seconds per cubic centimeter of iron is

$$W/V = \int_{C}^{A} (in/l) \cdot (d\Phi/S) = \int_{C}^{A} HdB, \quad \dots \quad (19)$$

where H is in ampere-turns per centimeter, and B is in webers per square centimeter.

But HdB is the area of an infinitesimal strip, such as is shown by hatching in Fig. 7.   Consequently, the right-hand side of eq.

(19) represents the area of the figure ACQ, which is therefore a measure for the energy transferred to the electric circuit, per cubic centimeter. In exactly the same way it can be shown that the energy supplied to the magnetic circuit during the part $C'A$ of the cycle is represented by the area $AC''Q$. Hence the *net* energy loss for the part of the cycle to the right of the axis of ordinates is represented by the area $ACC''A$. Repeating the same reasoning for the left-hand side of the loop it will be seen that *the total energy loss per cycle of magnetization per cubic centimeter of material is represented by the area $ACA'C''A$ of the hysteresis loop.* For a given material, this area, and consequently the loss, is a function of the maximum flux density PA, and increases with it according to a rather complicated law. Two empirical formulæ for the loss of energy as a function of the density are given in Art. 20 below.

In the problems that follow the weight of one cubic decimeter of solid carbon steel is taken to be 7.8 kg., and that of the alloyed or silicon steel 7.5 kg. The weight of one cubic decimeter of assembled carbon steel laminations is taken as $0.9 \times 7.8 = 7$ kg., and that of silicon steel laminations as $0.9 \times 7.5 =$ about 6.8 kg.

Prob. 5. A hysteresis loop is plotted to the following scales: abscissæ 1 cm. = 10 amp.-turns/cm.; ordinates, 1 cm. = 1 kilo-maxwell/sq. cm.; the area of the loop is found by a planimeter to be 72 sq. cm. What is the loss per cycle per cubic decimeter of iron?
                                        Ans. 7.2 watt-seconds (joules).
Prob. 6. The hysteresis loop mentioned in the preceding problem was obtained from an oscillographic record at a frequency of 60 cy., with a sample of iron which weighed 9.2 kg. What was the power lost in hysteresis in the whole ring?   Ans. 510 watts.
Prob. 7. The stationary coil of a ballistic electro-dynamometer is connected in series with the exciting electric circuit (Fig. 1); the moving coil is connected through a high resistance to a secondary winding placed on the ring. The exciting current is brought to a certain value, and then the current is reversed twice in rapid succession, in order that the iron may undergo a complete magnetization cycle. Show that the deflection of the electro-dynamometer is a measure for the area of the hysteresis loop. Hint: $HdB = H(dB/dt)\ dt = \text{Const.} \times ie\,dt.$[1]

17. Eddy Currents in Iron. Iron is an electrical conductor; therefore when a magnetic flux varies in it, electric currents are

---

[1] Searle, "The Ballistic Measurement of Hysteresis," *Electrician*, Vol. 49, 1902, p. 100.

induced along closed paths of least resistance linked with the flux. These currents permeate the whole bulk of the iron and are called eddy or Foucault currents. Eddy currents cause a loss of energy which must be supplied either electrically or mechanically from an outside source. Therefore, the iron cores used for variable fluxes are usually built of laminations, so as to limit the eddy currents to a small amount by interposing in their paths the insulation between the laminations. Japan, varnish, tissue paper, etc., are used for this purpose. In many cases the layer of oxide formed on laminations during the process of annealing is considered to be a sufficient insulation against eddy currents.

The usual thickness of lamination varies from 0.7 to 0.3 mm., according to the frequency for which an apparatus is designed, the flux density to be used, the provision for cooling, etc. The more a core is subdivided the lower is the loss due to eddy currents, but the more expensive is the core on account of the higher cost of rolling sheets, and of punching and assembling the laminations. Besides, more space is taken by insulation with thinner stampings, so that the per cent net cross-section of iron is reduced. The net cross-section of laminations is usually from 95 to 85 per cent of the gross cross-section, depending upon the thickness of the laminations, the kind of insulation used, and the care and pressure used in assembling the core. For preliminary calculations about ten per cent of the gross cross-section is assumed to be lost in insulation.

Fig. 9 shows two iron cores in cross-section, one core solid, the other subdivided into three laminations by planes parallel to the direction of the lines of force. The lines of force are shown by dots, and the paths of the eddy currents by continuous lines. Eddy currents are linked with the lines of force, the same as the current in the exciting winding. In fact, eddy currents are similar to the secondary currents in a transformer, inasmuch as they tend to reduce the flux created by the primary current. The core must be laminated in planes perpendicular to the lines of flow of the eddy currents, so as to break up their paths and at the same time not to interpose air-gaps in the paths of the lines of force.

An iron core can be further subdivided by using thin iron wires in place of laminations. Such cores were used in early machines and transformers, but were abandoned on account of

expense and poor space factor.   Iron-wire cores are used at present
in only high-frequency apparatus, in which eddy currents must
be carefully guarded against; for instance in the induction coils
(transformers) employed in telephone circuits.

It will be seen by an inspection of Fig. 9 that eddy currents
are much smaller in the laminated core because the resistance of
each lamination is increased while the flux per lamination and consequently the induced e.m.f. is considerably reduced.   It is proved
in Art. 21 below that the power lost in eddy currents per kilogram
of laminations is proportional to the square of the thickness of

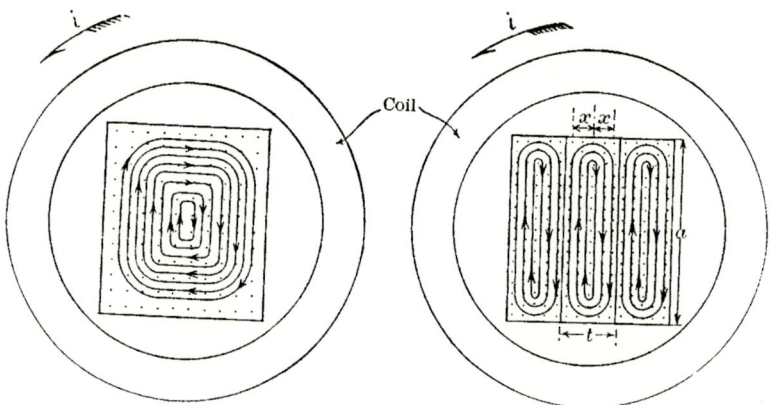

Fig. 9.—Eddy currents in a solid and in a laminated core.

the laminations, the square of the frequency, and the square of
the flux density.

Prob. 8. Show that the armature cores of revolving machinery must
be laminated in planes perpendicular to the axis of rotation.

Prob. 9. Show that assuming the temperature-resistance coefficient of
iron laminations to be 0.0046 per degree Centigrade the eddy current loss of
a core at 70° C. is only about 82.5 per cent of that at 20° C.

Prob. 10. Explain the reason for which the hysteresis loss in a given
core and at a given frequency depends only on the amplitude of the exciting current, while the eddy-current loss depends also upon the wave-form
of the current.

18. The Significance of Iron Loss in Electrical Machinery.
The power lost in an iron core on account of hysteresis and eddy
currents, taken together, is called *iron loss* or *core loss*.   It is of
importance to understand the effect of this loss in the iron cores

of electrical machinery and apparatus: First, because they bring about a loss of power and hence lower the efficiency of a machine; Secondly, because they heat up the iron and thus limit the permissible flux density, or make extra provisions for ventilation and cooling necessary; Thirdly, because they affect the indications of measuring instruments. The effects of hysteresis and eddy currents in the principal types of electrical machinery are as follows:

(a) In a transformer an alternating magnetization of the iron causes a core loss in it. The power thus lost must be supplied from the generating station in the form of an additional energy component of the primary current. The core is heated by hysteresis and by eddy currents, and the heat must be dissipated by the oil in which the transformer is immersed, or by an air blast.

(b) In a direct-current machine the revolving armature is subjected to a magnetization first in one direction and then in the other; the heating effect due to the hysteresis and eddy currents is particularly noticeable in the armature teeth in which the flux density is usually quite high. The core loss, being supplied mechanically, causes an additional resisting torque between the armature and the field. In a generator this torque is supplied by the prime mover; in a motor this torque reduces the available torque on the shaft.

(c) The effect of hysteresis and of eddy currents in the armature of an alternator or of a synchronous motor is similar to that in a direct-current machine.

(d) In an induction motor the core loss takes place chiefly in the stator iron and teeth, where the frequency of the magnetic cycles is equal to that of the power supply; the frequency in the rotor corresponds to the per cent slip, so that even with very high flux densities in the rotor teeth the core loss in the rotor is comparatively small. At speeds below synchronism the necessary power for supplying the iron loss is furnished electrically as part of the input into the stator. At speeds above synchronism this power is supplied through the rotor from the prime mover.

(e) In a direct-current ammeter, if it has a piece of iron as its moving element, residual magnetism in this iron causes inaccuracies in its indications. With the same current the indication of the instrument is smaller when the current is increasing than when it is decreasing; this can be understood with reference to the hysteresis loop.   With alternating current the effect of hysteresis is automatically eliminated by the reversals of the current which passes through the instrument.

From these examples the reader can judge as to the effect of hysteresis in other types of electrical apparatus not considered above.

Prob. 11. Show that in an 8-pole direct-current motor running at a speed of 525 r.p.m. the armature core and teeth undergo 35 complete hysteresis cycles per second.

Prob. 12. Show that for two points in an armature stamping, taken on the same radius, one in a tooth, the other near the inner periphery of the armature, the hysteresis loops are displaced in time by one-quarter of a cycle.

19. The Total Core Loss.   In practical calculations on electrical machinery the total core loss is of interest, rather than the hysteresis and the eddy current losses separately.   For such computations empirical curves are used, obtained from tests on steel of the same quality and thickness.   The curves of total core loss given in Fig. 10 have been compiled from various sources, and give a fair idea of the order of magnitude of core loss in various grades of commercial steel laminations.   The specimens were tested in the Epstein apparatus, which is a miniature transformer (see the author's *Experimental Electrical Engineering*, Vol. 1, p. 197), and the values given can be used for estimating the core loss in transformers and in other stationary apparatus with a simple magnetic circuit.

In using the curves one should note that the ordinates are watts per cubic decimeter of laminations, hence the gross volume and not the volume of the iron itself is represented.   On the other hand, the abscissæ are the true flux densities in the iron.   In choosing a material the following points are worthy of note: (1) Silicon steel is now used for 60-cycle transformers, almost to the exclusion of any other, on account of its lower core loss; it is sometimes used for 25-cycle transformers also.   (2) The material called " Good carbon steel " is that which is used for induction motor stators, and in general for the armatures of alternating and direct-current machinery; also, sometimes for the cores of·low frequency transformers.   (3) The material called " Ordinary carbon steel " should be used only in those cases for which the core loss is of small importance.

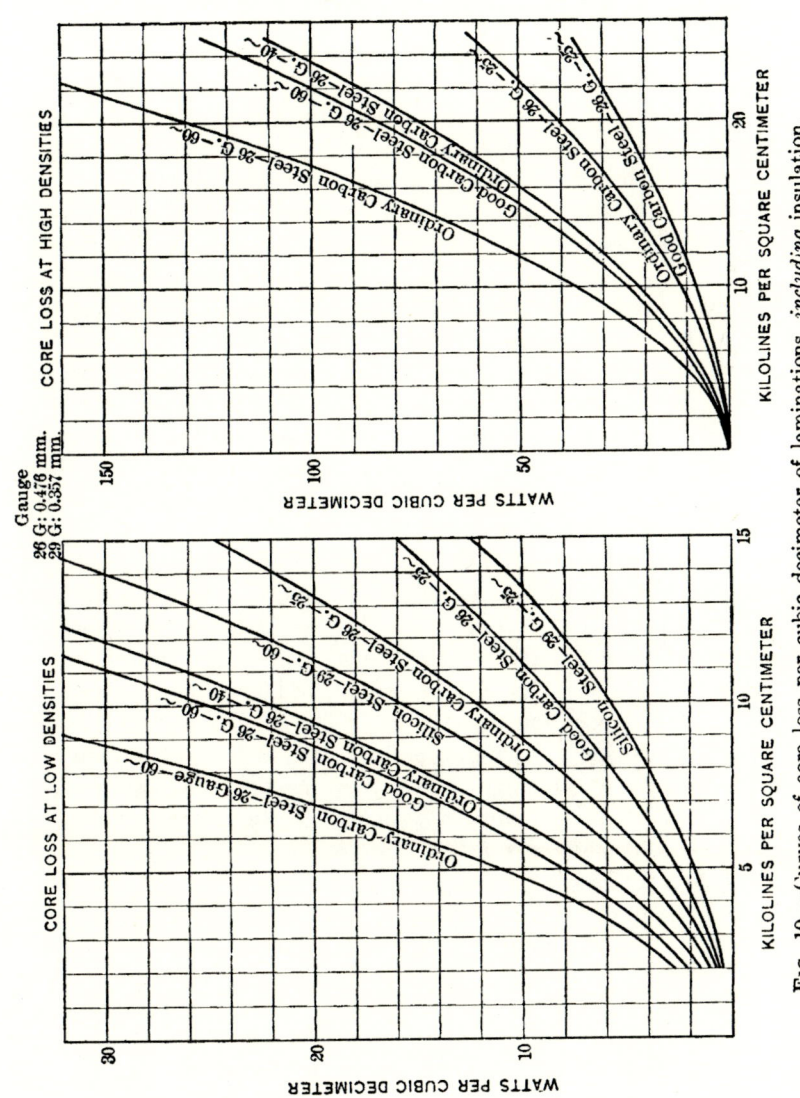

FIG. 10.—Curves of core loss per cubic decimeter of laminations, *including* insulation.

The thickness of lamination to be used in each particular case is a matter of judgment based on previous experience, and no general rule can be laid down, except what is said in Art. 17 above, in regard to the factors upon which the eddy-current loss depends. The gauges 26 to 29 are representative of the usual practice. If it should be necessary to estimate the core loss for a different thickness and at another frequency than those given in Fig. 10, the method explained in Art. 22 below may be used.

The core loss in the armatures and teeth of revolving machinery is found from tests to be considerably above that calculated from the curves of loss on the same material when tested in stationary strips. This is probably due in part to the fact that the conditions of magnetization are different in the two cases. In the one case the cycles of magnetization are due to a *pulsating* m.m.f., which simply changes its magnitude; in the other case to a *gliding* m.m.f., with which the magnetic intensity at a point changes its direction as well. Besides, the distribution of flux densities in teeth and in armature cores is very far from being uniform. Therefore, when using the curves given in Fig. 10, for the calculation of iron loss in generators and motors, it is necessary to multiply the results by certain empirical coefficients obtained from the results of tests made on similar machines. Mr. I. E. Hanssen recommends adding 30, 35, and 40 per cent to the loss calculated from the curves obtained on stationary samples when estimating the iron loss in an armature back of its teeth, at 25, 40, and 60 cycles respectively. For teeth he recommends adding 30, 60, and 80 per cent, at the same frequencies.[1] These values are quoted here merely to give a general idea of the magnitude of the excess of core loss in revolving machinery; a responsible designer should compile the values of such coefficients from actual tests made on the particular class of machines which he is designing.

Some engineers do not use for revolving machinery values of core loss obtained on stationary samples, but plot the curves of core loss obtained directly from tests on machines of a particular kind, for various frequencies and flux densities. This is a reliable and convenient method provided that sufficient data are available to separate the core loss in the teeth from that in the core itself. Mr. H. M. Hobart advocates this method, and curves of core loss

---

[1] Hanssen, "Calculation of Iron Losses in Dynamo-electric Machinery," *Trans. Amer. Inst. Elec. Eng.*, Vol. 28 (1909), Part II, p. 993.

obtained directly from actual machines will be found in his several books on electric machine design.

It is customary now to characterize a lot of steel laminations with respect to its core loss by the co-called *figure of loss* (Verlust-ziffer), which is the total core loss in watts per unit of weight, at a standard frequency and flux density. In Europe the figure of loss is understood to be the watts loss per kilogram of laminations, at 50 cycles and at a flux density of 10 kilolines per square centimeter; the test to be performed in an Epstein apparatus under definitely prescribed conditions.[1] Sometimes a second figure of loss is required, referring to a density of 15 kilolines per square centimeter, when the laminations are to be used at high flux densities. In this country a figure of loss is sometimes used which gives the watts loss per pound of material at 60 cycles and at a flux density of 60 kilolines per square inch (or else at 10 kilolines per square centimeter; see the paper mentioned in problem 20 below).

In some cases it is required to estimate the hysteresis and the eddy current losses separately; also it is sometimes necessary to separate the two losses knowing a curve of the total loss. These calculations are explained in the articles that follow.

Prob. 13. The core of a 60-cycle transformer weighs 89 kg.; the gross cross-section of the core is 8 by 10 cm., of which 10 per cent is taken by the insulation between the laminations. The total flux alternates between the values of ± 0.49 megaline. If the core is made of gauge 26 good carbon steel, what is the total core loss according to the curves in Fig. 10? Solution: The flux density is $490/(8 \times 10 \times 0.9) = 6.8$ kl/sq. cm. The core loss per cubic decimeter at this density and at 60 cycles is, according to the curve, equal 13 watts. The volume of the laminations, including the insulation, is $89/7 = 12.7$ cu. dm. The total loss is $13 \times 12.7 = 165$ watts.

Prob. 14. What flux density could be used in the preceding problem if the core were made of silicon-steel laminations, gauge 29, provided that the total core loss be kept the same in both cases?

Ans. About 9 kl/sq. cm.

Prob. 15. Calculate the core loss in the stationary armature of a 60-cycle 450-r.p.m. alternator of the following dimensions: bore 180 cm.; gross axial length 24 cm.; two air-ducts 0.8 cm. each; radial width of stampings back of the teeth, 15 cm.; the machine has 144 slots, 2 cm. wide and 4.5 cm. deep. The core is made of 26-gauge good carbon steel; the useful flux per pole is 4.65 megalines, and two-thirds of the total number of teeth carry the flux simultaneously. Use Mr. Hanssen's coefficients.

[1] See *Elektrotechnische Zeitschrift*, Vol. 24 (1903), p. 684.

Note: All parts of the core and all the teeth are subjected to complete cycles of magnetization in succession; therefore, in calculating the core loss the total volume of the core and of the teeth must be multiplied by the loss per cubic decimeter, corresponding to the maximum magnetic density in each part. The density in a tooth varies along its length, being a maximum at the tip. The average density may be assumed to be equal to that at the middle of the teeth.   Ans. About 9 kw.

20. Practical Data on Hysteresis Loss. The energy lost in hysteresis per cycle per kilogram of a given material depends only upon the maximum values of B and H, and does not depend upon the manner in which the magnetizing current is varied with the time between its positive and negative maxima. It is only at very high frequencies, such as are used in wireless telegraphy, that the particles of iron do not seem to be able to follow in their grouping the corresponding changes in the exciting current. With such high frequencies iron cores are not only useless, but positively harmful. However, at ordinary commercial frequencies the loss of power P_h due to hysteresis is proportional to the number of cycles per second and can be expressed as

$$P_h = f \cdot V \cdot F(B) \text{ watt.,}$$

where f is the number of magnetic cycles per second, V is the volume of the iron, and $F(B)$ is a function of the maximum flux density B. $F(B)$ represents the loss per cycle per cubic unit of material, and is therefore equal to the area of the hysteresis loop in Fig. 7.

One can assume empirically that the unit loss per cycle, $F(B)$, increases as a certain power n of B, this power to be determined from tests. The preceding formula becomes then

$$P_h = \eta f V B^n \text{ watt.,} \ldots \ldots (20)$$

where η is an empirical coefficient which depends upon the quality of the iron and upon the units used. Dr. Steinmetz found from numerous experiments that the exponent n varies between 1.5 and 1.7, and proposed for practical use the formula

$$P_h = \eta f V B^{1.6} \times 10^{-7} \text{ watt,} \ldots \ldots (21)$$

where the factor 10^{-7} is introduced in order to obtain convenient values for η when B is in maxwells per square centimeter, and V is in cubic centimeters. It is more convenient for practical calculations to use B in kilolines per square centimeter, and V in cubic decimeters. In this case the constant 10^{-7} is not necessary (see Prob. 17 below); but the student must now remember to multiply by 6.31 the values of η found in the various pocketbooks.

Hysteresis loss cannot be represented always with sufficient accuracy by formula (21) or (20) over a wide range of values of B, because the exponent n itself seems to increase with B. Where greater accuracy is required at medium and high flux densities the following formula may be used:

$$P_h = fV(\eta'B + \eta''B^2). \quad \cdots \quad (21a)$$

In this formula the term with B^2 automatically becomes of more and more importance as B increases. By selecting proper values for η' and η'' a given experimental curve of loss can be approximated more closely than by means of formula (21). On the other hand, formula (21) is more convenient for comparison and analysis.

Curves of hysteresis loss and values of the constant η will be found in various handbooks and pocketbooks. It is hardly worth while giving them here, because hysteresis loss varies greatly with the quality of iron and with the treatment it is given before use. Moreover, the quality of the iron used in electrical machinery is being improved all the time, so that a value of η given now may be too large a few years from now.

Considerable effort is being constantly made to improve the quality of the iron used in electrical machinery so as to reduce its hysteresis loss. The latest achievement in this respect is the production of the so-called *silicon steel*, also called *alloyed steel*, which contains from 2.5 to 4 per cent of silicon. This steel shows a much lower hysteresis loss than ordinary carbon steel. Incidentally, the electric resistivity of silicon steel is about three times higher than that of ordinary steel, so that the eddy-current loss is reduced about three times. The advantage that silicon steel has over carbon steel is clearly seen in Fig. 10. Silicon steel is largely used for transformer cores because it permits the use of higher flux densities, and therefore the reduction of the weight and cost of a transformer, in spite of the fact that silicon steel itself costs more per kilogram than carbon steel.

Another great advantage of silicon steel is that it is practically *non-aging*; this means that the hysteresis loss does not increase with time. An increase in the hysteresis loss of a transformer during the first few years of its operation used to be a serious matter in the design and operation of transformers, because of the subsequent overheating of the core and of the coils. Silicon steel shows practically no increase in its hysteresis loss after several years of operation. Moderate heating, which considerably increases the hysteresis loss in ordinary steel, has no effect on silicon steel.

Impurities which are of such a nature as to produce a softer iron or steel and a material of higher permeability, are as a rule favorable to the reduction of the hysteresis loss, and *rice versa*. Mechanical treatment and heating are also very important in their effects on hysteresis loss. In particular, punching and hammering increases hysteresis loss, while annealing reduces it. Therefore laminations are always annealed carefully after being punched into their final shape.

The requirements for the steel used in permanent magnets are entirely different from those for the cores of electrical machinery. In permanent magnets a large and wide hysteresis loop is desired, because it means a high percentage of residual magnetism (ratio of CO to AP, Fig. 7) and a large coercive force, OF. Both are favorable for obtaining strong permanent magnets of lasting strength. Combined carbon is particularly important for obtaining these qualities, as is also the proper heat treatment after magnetization.

Prob. 16. In the 60-cycle transformer given in prob. 13, the core weighs 89 kg. and is made of 26 gauge good carbon steel. The maximum flux density is 6.8 kl./sq. cm. What is the hysteresis loss assuming η to be equal to 0.0012? Ans. About 124 watt.

Prob. 17. What is the constant in formula (21) in place of 10^{-7}, if, with the same η, the density B is in kilo-maxwells per sq. cm., and the volume is in cubic decimeters? Ans. 6.31.

Prob. 18. Show how to determine the values of η and n in eq. (20), knowing the values W_1 and W_2 of the energy lost per cycle at two given values of maximum flux density, B_1 and B_2.
Ans. $n = (\log W_2 - \log W_1)/(\log B_2 - \log B_1)$.

Prob. 19. The following values of hysteresis loss per cu. decimeter have been determined from a test at 25 cycles (after eliminating the eddy current loss):

Flux density in kl/sq.cm., $B =$ 5.0	6.5	8.0	10.0
Hysteresis loss in watts, $P_h =$ 1.30	2.00	2.88	4.11

What are the values of η and n in formula (20)? Suggestion: Use logarithmic paper to determine the most probable value of n, by drawing the straight line log $P_h = n$ log B + log Const.   See the author's *Experimental Electrical Engineering*, Vol., 1, p. 202.

Ans. $P_h = 0.00368 \ fVB^{1.65}$.

21. Eddy Current Loss in Iron.   With the thin laminations used in the cores of electrical machinery the eddy-current loss in watts can be represented by the formula

$$P_e = \varepsilon V(tfB)^2, \ . \ . \ . \ . \ . \ . \ . \quad (22)$$

where ε is a constant which depends upon the electrical resistivity of the iron, its temperature, the distribution of the flux, the wave form of the exciting current, and the units used. V is the volume or the weight of the core for which the loss is to be computed; t is the thickness of laminations, f the frequency of the supply, and B the maximum flux density during a cycle.   If B is different at different places in the same core, the average of these should be taken, (B is the time maximum but the space average).   Sometimes formula (22) contains also 10 to some negative power in order to obtain a convenient value of ε.

Formula (22) can be proved as follows: The loss of power in a lamination can be represented as a sum of the i^2r losses for the small filaments of eddy current in it.   But $i^2r = e^2/r$; it can be shown that the expression in parentheses in formula (22) is proportional to the sum of e^2/r per unit volume. When the frequency f increases say n times, the rate of change of the flux, $d\Phi/dt$, and consequently the e.m.fs. induced in the iron are also increased n times.   Therefore, the loss which is proportional to e^2 increases n^2 times.   In other words, the loss is proportional to the square of the frequency.   Similarly, the induced voltage is proportional to the flux density B; and consequently, the loss is proportional to B^2.

To prove that the loss is proportional to the square of the thickness of laminations one must remember that increasing the thickness n times increases the flux and the induced e.m.f. within any filament of eddy current also n times.   But the resistance of each path is reduced n times (neglecting the short sides of the rectangle).   Consequently, the expression e^2/r is increased n^3 times. However, inasmuch as the volume of the lamination is also creased n times, the loss *per unit volume* is only n^2 times larger.   In other words, the loss per unit volume increases as t^2.   A more rigid proof of this proposition is given in problem 21 below.

For values of ε the reader is referred to pocketbooks; the numerical values given there must, however, be used cautiously, because the eddy-current loss depends on some factors such as the care exercised in assembling, and the actual distribution of the flux, which factors can hardly be taken into account in a formula. As a matter of fact, formula (22) is used now less and less in practical calculations, the engineer relying more upon experimental curves of *total* core loss (Fig. 10).

Prob. 20. According to the experiments of Lloyd and Fischer [*Trans-Amer. Inst. Elec. Engs.*, Vol. 28 (1909), p. 465] the eddy-current loss in silicon-steel laminations of gauge 29 (0.357 mm. thick) is from 0.12 to 0.18 watts per pound at 60 cycles and at $B = 10,000$ maxwells per sq. cm. What is the value of the coefficient ε in formula (22) if P is in microwatts, V is the weight of the core in kg. (not the volume, as before); if also t is in mm., and B is in kilolines per sq. cm.?
    Ans. From 5.78 to 8.67; 7.2 is a good practical average.

Prob. 21. Prove that the loss of power caused by eddy currents, per unit volume of thin laminations, is proportional to the square of the thickness of the laminations. Solution: The thickness t of the sheet (Fig. 9) being by assumption very small as compared with its width a, the paths of the eddy current may be considered to be rectangles of the length a and of different widths, ranging from t to zero. Consider one of the tubes of flow of current, of a width $2x$, thickness dx, and length l in the direction of the lines of magnetic force. Let the flux density vary with the time between the limits $\pm B$. Then the maximum flux linking with the tube of current under consideration is approximately equal to $2axB$; therefore, the effective value of the voltage induced in the tube can be written in the form $e = CaxBf$, where C is a constant, the value of which we are not concerned with here. The resistance of the tube is $\rho(2a + 4x)/(ldx)$, or very nearly $2a\rho/(ldx)$. Thus we have that the i^2r loss, or the value of e^2/r for the tube under consideration, is $dP_e = C^2ax^2B^2f^2ldx/2\rho$. Integrating this expression between the limits 0 and $t/2$ we get $P_e = C^2at^3B^2f^2l/48\rho$. But the volume of the lamination is $V = atl$. Dividing P by V we find that the loss per unit volume is proportional to $(tfB)^2$.[1]

Prob. 22. Prove that the loss of power by eddy currents per unit volume in round iron wires is proportional to the square of the diameter of the wire. The flux is supposed to pulsate in the direction of the axes of the wires, and the lines of flow of the eddy currents are concentric circles. Hint: Use the method employed in the preceding problem.

22. The Separation of Hysteresis from Eddy Currents. It is
sometimes required to estimate the total core loss for a thickness

[1] For a complete solution of this and the following problem, including the numerical values of C, see Steinmetz, *Alternating Current Phenomena* (1908), Chap. XIV.

of steel laminations other than those given in Fig. 10, or at a different frequency.   For this purpose, it is necessary to separate the loss due to hysteresis from that due to eddy currents, because the two losses follow different laws, expressed by eqs. (20) and (22) respectively.

In order to separate these losses at a certain flux density it is necessary to know the value of the total core loss at this density, and at two different frequencies.   For a given sample of laminations, the total core loss P at a constant flux density and at a variable frequency f, can be represented in the form

$$P = Hf + Ff^2, \quad \ldots \ldots \quad (23a)$$

where Hf represents the hysteresis loss, and Ff^2 the eddy or Foucault current loss. H is the hysteresis loss per cycle, and F is the eddy-current loss when f is equal to one cycle per second.   Writing this equation for two known frequencies, two simultaneous equations are obtained for H and F, from which H and F can be determined.

In practice the preceding equation is usually divided by f, because in the form

$$P/f = H + Ff, \quad \ldots \ldots \ldots \quad (23b)$$

it represents the equation of a straight line between P/f and f. This form is particularly convenient when the values of P are known for more than two frequencies.   In this case the values of P/f are plotted against f as abscissæ, and the most probable straight line is drawn through the points thus obtained.   The intersection of this straight line with the axis of ordinates gives directly the value of H.   After this, F is found from eq. (23b).

Knowing H and F at a certain flux density, the separate losses Hf and Ff^2 can be calculated for any desired frequency.   For the same material, but of a different thickness, the hysteresis loss per kilogram weight is the same, while the eddy-current constant F varies as the square of the thickness, according to eq. (22).   Thus, knowing the eddy loss at one thickness it can be estimated for any other thickness.

It is sometimes required to estimate the iron loss at a flux density higher than the range of the available curves; in other words, the problem is sometimes put to extrapolate a curve like one of those in Fig. 10.   There are two cases to be considered.

(*A*) If two or more curves for the same material are available, taken at different frequencies, the hysteresis is first separated from the eddy-current loss as is explained before, for several flux densities within the range of the curves. Then the exponent, according to which the hysteresis loss varies with the flux density is found, by plotting the hysteresis loss to a logarithmic scale (see problems 18 and 19 above). Finally the two losses are extrapolated. In extrapolating, the hysteresis loss is assumed to vary according to the same law, and the eddy current loss is assumed to vary as the square of the flux density; see eq. (22).

(*B*) Should only one curve of the total loss be available for extrapolation, this curve may be assumed to be a parabola of the form $P = aB + bB^2$. Dividing the equation throughout by B we get

$$P/B = a + bB. \quad \ldots \ldots \quad (24)$$

This is the equation of a straight line between P/B and B. Plotting P/B against the values of B as abscissæ, a straight line is obtained which can be easily extrapolated. In some cases the values of P/B thus plotted give a line with a perceptible curvature. Nevertheless, the curvature is much smaller than that of the original P curve, so that the P/B curve can be extrapolated with more certainty, especially if the lower points be disregarded.[1]

Prob. 23. From the curves in Fig. 10 calculate the core loss per cubic decimeter of 29-gauge silicon-steel laminations, at a flux density of 10 kl/sq.cm. and at 40 cycles.   Ans.   About 10 watts.

Prob. 24. Using the data obtained in the solution of the preceding problem calculate the figure of loss of 26-gauge laminations at 60 cycles.   Ans.   2.7 watt/kg.

Prob. 25. Check the curve of total core loss for the ordinary carbon steel at 40 cycles with the curves for 25 and 60 cycles.

Prob. 26. Extrapolate the curve of core loss for the silicon steel at 25 cycles up to the density of 20 kl./sq.cm. Which of the two methods is preferable?   Ans.   21 watts per cu.dm. at $B = 20$.

Prob. 27. Show that the core loss curve for ordinary carbon steel, at 60 cycles, follows closely eq. (24).

[1] If the P/B curve should prove to be a straight line, then it is probable that the hysteresis loss follows eq. 21a more nearly than eq. 20. In this case, even if we had data for two frequencies, method (*B*) would be both more accurate, and more simple than method (*A*).

CHAPTER IV

INDUCED E.M.F. IN ELECTRICAL MACHINERY

23. Methods of Inducing E.M.F. The following are the principal cases of induced e.m.f. in electrical machinery and apparatus:

(a) In a transformer, an alternating magnetizing current in the primary winding produces an alternating flux which links with both windings and induces in them alternating e.m.fs. A similar case is that of a variable current in a transmission line which induces a voltage in a telephone line which runs parallel to it.

(b) In a direct-current machine, in a rotary converter, and in a homopolar machine electromotive forces are induced in the armature conductors by moving them across a stationary magnetic field.

(c) In an alternator and in a synchronous motor, with a stationary armature and a revolving field, electromotive forces are induced by making the magnetic flux travel past the armature conductors.

(d) In a polyphase induction motor the currents in the stator and in the rotor produce together a resultant magnetomotive force which moves along the air-gap and excites a gliding (revolving) flux. This flux induces voltages in both the primary and the secondary windings.

(e) In a single-phase motor, with or without a commutator, the e.m.fs. induced in the armature are partly due to the " transformer action," as under (a), and partly to the " generator action," as under (b).

(f) In an inductor-type alternator both the exciting and the armature windings are stationary; the pole pieces alone revolve. The flux linked with the armature coils varies periodically, due to the varying reluctance of the magnetic circuit, because of the motion of the pole pieces. This varying flux induces an alternating e.m.f. in the armature winding. Or else, one may say that the flux travels along the air-gap with the projecting poles, and cuts the armature conductors.

(g) Whenever the current varies in a conductor, e.m.fs. are induced not only in surrounding conductors but also in the conductor itself. This e.m.f. is called the e.m.f. of self-induction. Such e.m.fs. are present in alternating-current transmission lines, in the armature windings of alternating-current machinery, etc. While the e.m.f. of self-induction does not differ fundamentally from the transformer action mentioned above, its practical aspect is such as to make a somewhat different treatment desirable. Inductance and its effects are therefore considered separately in chapters X to XII below.

All of the foregoing cases can be reduced to the following two fundamental modes of action of a magnetic flux upon an electrical conductor:

(1) The exciting magnetomotive force and the winding in which an e.m.f. is to be induced are both stationary, relatively to one another; in this case the voltage is induced by a varying magnetic flux. Changes in the flux are produced by varying either the magnitude of the m.m.f., or the reluctance of the magnetic circuit. This method of inducing an e.m.f. is usually called the *transformer action*.

(2) The exciting magnetomotive force and the winding in which the e.m.f. is to be induced are made to move relatively to each other, so that the armature conductors *cut* across the lines of the flux. This method of inducing an e.m.f. is conventionally referred to as the *generator action*.

By analyzing the transformer action more closely it can be reduced to the generator action, that is to say to the " cutting " of the secondary conductor by lines of magnetic flux or force. This is so, because in reality the magnetic disturbance spreads out in all directions from the exciting winding, and when the current in the exciting winding varies the magnetic disturbance travels to or from the winding in directions perpendicular to the lines of force (Fig. 11). This traveling flux cuts the secondary conductor and induces in it an e.m.f. However, the question as to whether an e.m.f. is induced by a change in the total flux within a loop, or by the cutting of a conductor by a magnetic flux is still in a somewhat controversial state;[1] although Hering's experiment is a strong

[1] Carl Hering, "An Imperfection in the Usual Statement of the Funda-

argument in favor of the theory of "cutting" of lines of force. He showed that no e.m.f. is induced in an electric circuit when a flux is brought in or out of it without actually cutting any of the conductors of the electric circuit. For practical purposes it is convenient to distinguish the transformer action from the generator action, so that the matter of unifying the statements (1) and (2) into one more general law is of no immediate importance.

24. The Formulæ for Induced E.M.F.   In accordance with the definition of the weber given in Art. 3, we have

$$e = -d\Phi/dt, \quad \ldots \ldots \ldots \quad (25)$$

where e is the instantaneous e.m.f. in volts, induced by the trans-

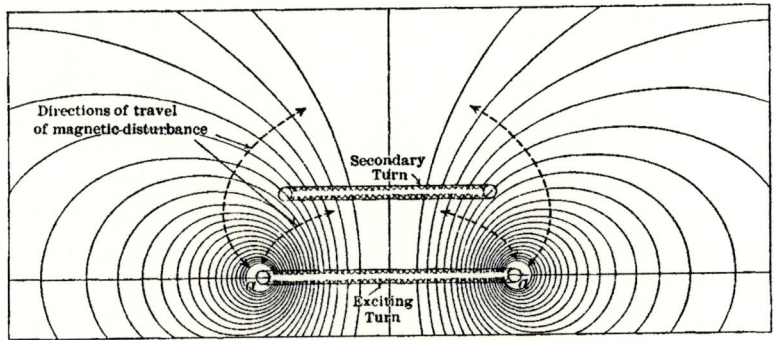

Fig. 11.—E.M.F. induced by transformer action.

former action in a turn of wire which at the time t is linked with a flux of Φ webers.   The value of e is determined not by the value of Φ but by the rate at which Φ varies with the time.   In the case of the generator action $d\Phi$ in formula (25) represents the flux which the conductor under consideration cuts during the interval of time dt. It can be shown that the two interpretations of $d\Phi$ lead to the same result.   Namely, in the case of the transformer action (Fig. 11), the new flux, $d\Phi$, is brought within the secondary turn by cutting through the conductor of this turn.   Therefore, in the case of the

mental Law of Electromagnetic Induction, *Trans. Amer. Inst. Elec. Engs.*, Vol. 27 (1908), Part. 2, p. 1341.   Fritz Emde, Das Induktionsgesetz, *Elektrotechnik und Maschinenbau*, Vol. 26 (1908); Zum Induktionsgesetz, *ibid.*, Vol. 27 (1909); De Baillehache, Sur la Loi de l'Induction, *Bull. Societé Internationale des Electriciens*, Vol. 10 (1910), pp. 89 and 288.

transformer action $d\Phi$ can also be considered as the flux which cuts the loop during the time dt, the same as in the generator action. On the other hand, the moving conductor in a generator is a part of a turn of wire, and any flux which it cuts either increases or decreases the total flux linking with the loop. Consequently, in the case of the generator action $d\Phi$ can be interpreted as the change of flux within the loop, the same as in the transformer action. Thus, the mathematical expression for the induced e.m.f. is the same in both cases, provided that the proper interpretation is given to the value of $d\Phi$.

The sign minus in formula (25) is understood with reference to the right-hand screw rule (Art. 1), i.e., with reference to the direction of the current which would flow as a result of the induced electromotive force. Namely, the law of the conservation of energy requires that this induced current must oppose any change in the flux linking with the secondary circuit. If this were otherwise, a slight increase in the flux would result in a further indefinite increase in the flux, and any slight motion of a conductor across a magnetic field would help further motion.

The positive direction of the induced e.m.f. is understood to be that of the primary current which excites the flux at the moment under consideration. If the flux linked with the secondary circuit increases, $d\Phi/dt$ in formula (25) is positive, but the secondary current must be opposite to the primary in order to oppose the increase. Thus, the secondary current is negative, and by assumption the induced e.m.f. e is also negative. Therefore, the sign minus is necessary in the formula. When the flux decreases, $d\Phi/dt$ is negative, but the secondary current is positive, because it must oppose the reduction in flux. Hence, in order to make e a positive quantity, the sign minus is again necessary.

The following two special cases of formula (25) are convenient in applications. Formula (25) gives the instantaneous value of the induced e.m.f.; it is in some cases required to know the average e.m.f. induced during a finite change of the flux from Φ_1 to Φ_2. By definition, the average e.m.f. is

$$e_{arc} = \frac{1}{t_2 - t_1} \int_{t_1}^{t_2} e\, dt,$$

where t_1 is the initial moment and t_2 the final moment of the interval of time during which the change in the flux takes place.

Substituting in this equation the value of e from (25), and integrating, we get

$$e_{ave} = (\Phi_1 - \Phi_2)/(t_2 - t_1). \quad . \quad . \quad . \quad . \quad (26)$$

This shows that the average value of an induced e.m.f. does not depend upon the law according to which the flux changes with the time, and is simply proportional to the average rate of change of the flux.

As another special form of eq. (25) consider a straight conductor of a length l centimeters moving at a velocity of v centimeters per second across a uniform magnetic field of a density of B webers per sq. cm. Let B, l, and v be in three mutually perpendicular directions. The flux $d\Phi$ cut by the conductor during an infinitesimal element of time dt is equal to $Blv\,dt$. Substituting this value into eq. (25) we get, apart from the sign minus,

$$e = Blv. \quad . \quad . \quad . \quad . \quad . \quad . \quad (27)$$

Should the three directions, B, l, and v, be not perpendicular to each other, l in eq. (27) is understood to mean the projection of the actual length of the conductor, perpendicular to the field, and v is the component of the velocity normal to B and l. Both B and v may vary with the position of the conductor, in which case eq. (27) gives the value of the instantaneous voltage. If, at a certain moment, the various parts of the conductor cut across a field of different density, eq. (27) must be written for an infinitesimal length of the conductor, thus: $de = Bv \cdot dl$, and integrated over the whole length of the conductor.

Besides the rule given above, the direction of the e.m.f. induced by the generator action can also be determined by the familiar three-finger rule, due to Fleming, and given in handbooks and elementary books on electricity. This rule is useful beause it emphasizes the three mutually perpendicular directions, those of the flux, the conductor, and the relative motion. In applying this rule to a machine with a stationary armature one must remember that the direction of the motion in Fleming's rule is that of the conductor, and therefore is opposite to the direction of the actual motion of the magnetic field.

Problem 1. A secondary winding is placed on the ring (Fig. 1) and is connected to a ballistic galvanometer. Let the number of turns in the secondary winding be n, the flux linking with each turn be Φ webers, and the total resistance of the secondary circuit be r ohms. Show that when the current in the primary winding is reversed, the discharge through the galvanometer is equal to $2n\phi/r$, in coulombs.

Prob. 2. A telephone line runs parallel to a direct-current trolley feeder for 20 kilometers. When a current of 100 amperes flows through the feeder a flux of 2 kilo-maxwells threads through the telephone loop, per meter of its length. What is the average voltage induced in the telephone line when the current in the trolley feeder drops from 600 to 50 amp. within 0.1 sec.? Ans. 22 volts.

Prob. 3. Determine the number of armature conductors in series in a 550 volt homopolar generator of the axial type, running at a peripheral speed of about 100 meters per sec., when the length of the armature iron is 50 centimeters, and the flux density in the air-gap is between 18 and 19 kilolines per sq. cm. *Note:* For the construction of the machine see the *Standard Handbook*, index, under "Generators, homopolar."

Ans. 6.

Prob. 4. Draw schematically the armature and the field windings of a shunt-wound direct-current generator, select a direction of rotation, and show how to connect the field leads to the brushes so that the machine will excite itself in the proper direction.

Prob. 5. From a given drawing of a direct-current motor predict its direction of rotation.

Prob. 6. In an interpole machine the average reactance voltage per commutator segment during the reversal of the current is calculated to be equal to 34 volts. What is the required net axial length of the commutating pole to compensate for this voltage if the peripheral speed of the machine is 65 meters per second, and the flux density under the pole is 6 kl./sq.cm.? The armature winding has two turns per commutator segment. Ans. 22 cm.

25. The Induced E.M.F. in a Transformer.

The three types of transformers used in practice are shown in Figs. 12, 13, and 14. Considering the iron core as a magnetic link, and a set of primary and secondary coils as an electric link, one may say that the core-type transformer has one magnetic link and two electric links; the shell-type has one electric link and two magnetic links; the combination or cruciform type has one electric and four magnetic links. Still another type, not used in practice, can be obtained from the core-type by adding two or more electric links to the same magnetic link. Each electric link is understood to consist of two windings: the primary and the secondary.

When the primary winding is connected to a source of constant-potential alternating voltage and the secondary winding is connected to a load, alternating currents flow in both windings and an alternating magnetic flux is established in the iron core. If the primary electric circuit, that is, the one connected to the source of power, were perfect, that is, if it possessed no resistance and no reactance, the alternating magnetic flux in the core would be the same at all loads. It would have such a magnitude that at any instant the counter-e.m.f. induced by it in the primary winding would be practically equal and opposite to the impressed voltage. In reality the resistance and the leakage reactance of ordinary commercial transformers are so low that *for the purposes of calculating the magnetic circuit* the primary impedance drop may be disregarded, and the magnetic flux considered constant and independent of the load.

If the primary applied voltage varies according to the sine law, which condition is nearly fulfilled in ordinary cases, the counter-e.m.f., which is practically equal and opposite to it, also follows the same law. Hence, according to eq. (25), the magnetic flux must vary according to the cosine law, because the derivative of the cosine is minus the sine. In other words, both the flux and the induced e.m.f. vary according to the sine law, but the two

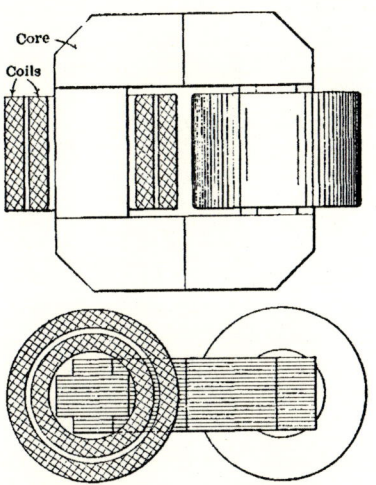

Fig. 12.—A core-type transformer.

sine waves are in time quadrature with each other. When the flux reaches its maximum its rate of change is zero, and therefore the counter-e.m.f. is zero. When the flux passes through zero its rate of change with the time is a maximum, and therefore the induced voltage at this instant is a maximum.

Let Φ_m be the maximum value of the flux in the core, in webers, and let f be the frequency of the supply in cycles per second. Then the flux at any instant t is $\Phi = \Phi_m \cos 2\pi ft$, and the e.m.f. induced at this moment, *per turn* of the primary or secondary winding is

$$e = -d\Phi/dt = 2\pi f\Phi_m \sin 2\pi ft.$$

Thus, the maximum value of the induced voltage per turn is

$2\pi f \Phi_m$; hence the effective value is $2\pi f \Phi_m / \sqrt{2} = 4.44 f \Phi_m$. Let there be N_1 primary turns in series; the total primary voltage is then equal to N_1 times the preceding value. Expressing the flux in megalines we therefore obtain the following practical formula for the induced voltage in a transformer:

$$E_1 = 4.44 f N_1 \Phi_m 10^{-2} \quad \ldots \ldots \quad (28)$$

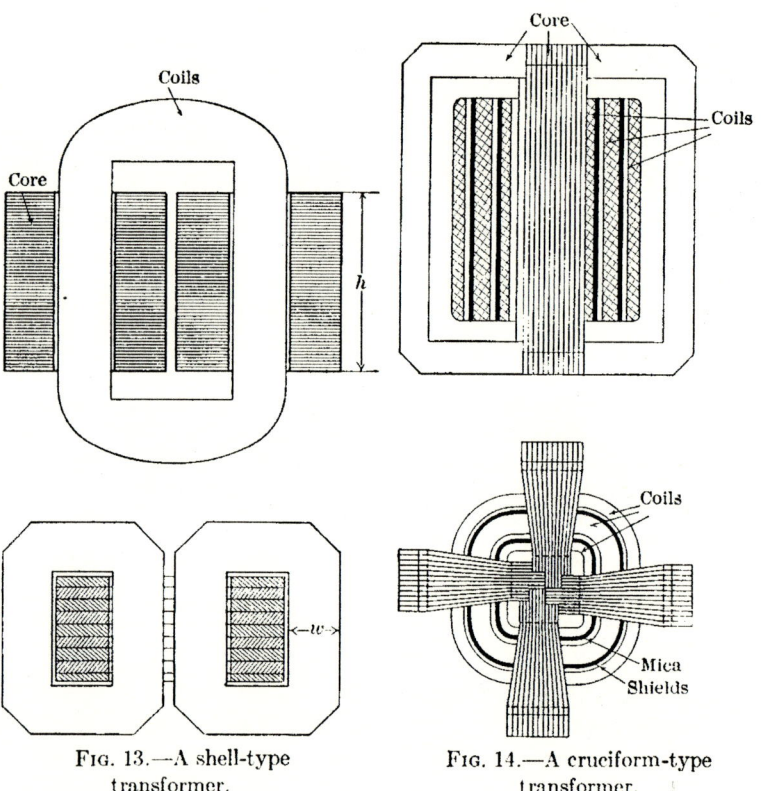

FIG. 13.—A shell-type   FIG. 14.—A cruciform-type
    transformer.   transformer.

In practice, E_1 is assumed to be equal and opposite to the applied voltage (for calculating the flux only, but not for determining the voltage regulation of the transformer). Formula (28) holds also for the secondary induced voltage E_2 if the number of secondary turns in series N_2 be substituted for N_1. The voltage per turn is the same in the primary and in the secondary winding; therefore, the ratio of the induced voltages is equal to that of the number

of turns in the primary and secondary windings: that is, we
have $E_1:E_2 = N_1:N_2$.

Prob. 7. A 60-cycle transformer is to be designed so as to have a flux
density in the core of about 9 kl./sq.cm.; the difference of potential
between consecutive turns must not exceed 5 volts. What is the required
cross-section of the iron?   Ans.   210 sq.cm.

Prob. 8. The transformer in the preceding problem is to be wound for
6600 v. primary, and 440 v. secondary. What are the required numbers
of turns?   Ans.   1320 and 88.

Prob. 9. Referring to the transformer in the preceding problem, what
are the required numbers of turns if three such transformers are to be used
Y-connected on a three-phase system, for which the line voltages are 6600
and 440 respectively?   Ans.   765 and 51.

Prob. 10. In a 110-kilovolt, 25-cycle transformer for Y-connection the
net cross-section of the iron is about 820 sq.cm. and the permissible maximum flux density is about 10.7 kl/sq.cm. What is the number of turns
in the high-tension winding?   Ans.   6500.

Prob. 11. The secondary of the transformer in the preceding problem
is to be wound for 6600 v., delta connection, with taps for varying the
secondary voltage within ±5 per cent. Specify the winding.
           Ans.   709 turns; taps taken after the 34th and 68th turn.

Prob. 12. Explain the reason for which a 60-cycle transformer usually
runs hot even at no load, when connected to a 25-cycle circuit of the same
voltage. Show from the core-loss curves that the voltage must be
reduced to from 75 to 85 per cent of its rated value in order to have the
normal temperature rise in the transformer, at the rated current.

Prob. 13. Show graphically that the wave of the flux, within a transformer, becomes more and more peaked when the wave of the applied e.m.f.
becomes more and more flat, and vice versa. Hint: The instantaneous
values of e.m.f. are proportional to the values of the slope of the curve of
flux.

Prob. 14. The wave of the voltage impressed upon a transformer has
a 15 per cent third harmonic which flattens the wave symmetrically. Show
analytically that the corresponding flux wave has a 5 per cent third harmonic in such a phase position as to make the flux wave peaked.

26. The Induced E.M.F. in an Alternator and in an Induction
Motor. Part of a revolving field alternator is shown in Fig. 15.
The armature core is stationary and has a winding placed in slots,
which may be either open or half closed. The pole pieces are
mounted on a spider and are provided with an exciting winding.
When the spider is driven by a prime mover the magnetic flux
sweeps past the armature conductors and induces alternating
voltages in them.[1]   In order to obtain an e.m.f. approaching a sine

[1] For details concerning the different types of armature windings see the
author's *Experimental Electrical Engineering*, Vol. 2, Chap. 30.

wave as nearly as possible the pole shoes are shaped as shown in the sketch, that is to say, so as to make a variable air-gap and thus grade the flux density from the center of the pole to the edges. In high-speed turbo-alternators the field structure often has a smooth surface, without projecting poles (Fig. 33), in order to reduce the noise and the windage loss. Such a structure is also stronger mechanically than one with projecting poles. The grading of the flux is secured by distributing the field winding in slots, so that the whole m.m.f. acts on only part of the pole pitch.

Consider a conductor at a during the interval of time during which the flux moves by one pole pitch τ. The average e.m.f.

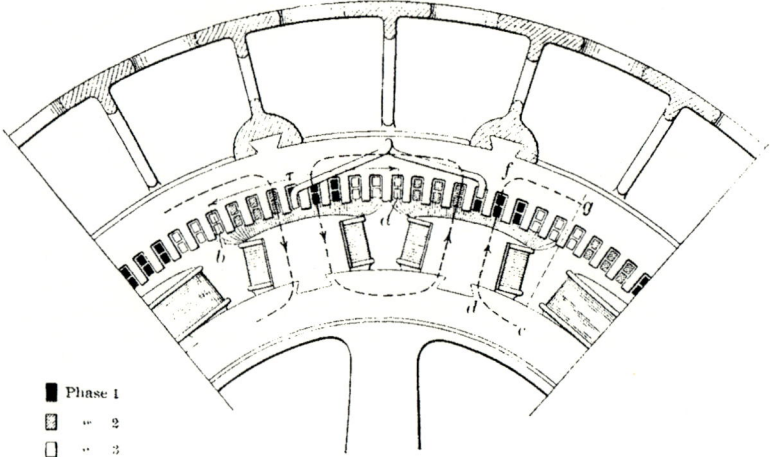

Phase 1
" 2
" 3

Fig. 15.—The cross-section of a synchronous machine.

induced in the conductor is, according to eq. (26), equal to $\Phi/\frac{1}{2}T$. where Φ is the total flux per pole in webers, and T is the time of one complete cycle, corresponding to 2τ the space of two pole pitches. But $T = 1/f$, so that the average voltage induced in a conductor is

$$e_{ave} = 2f\Phi. \qquad \ldots \ldots \qquad (29)$$

The value of e_{ave} thus does not depend upon the distribution of the flux Φ in the air-gap.

If the pole-pieces are shaped so as to give an approximately sinusoidal distribution of flux in the air-gap, the induced e.m.f. is also approximately a sine wave, and the ratio between the effective and the average values of the voltages is equal to $\frac{1}{2}\pi/\sqrt{2}$ or 1.11.[1] If the shape of the induced e.m.f. departs widely from the sine wave the actual curve must be plotted and its form factor determined by one of the known methods (see the *Electric Circuit*). Let the form factor in general be χ and let the machine have N armature turns in series per phase, or what is the same, $2N$ conductors in series. The total induced e.m.f. in effective volts is then

$$E = 2f\chi\Phi 2N. \quad . \quad . \quad . \quad . \quad . \quad (30)$$

This formula presupposes that there is but one slot per pole per phase, so that the e.m.fs. induced in the separate conductors are all in phase with each other, and that their values are simply added together. In reality, there is usually more than one slot per pole per phase, for practical reasons discussed in the next article. It will be seen from the figure that the e.m.fs. induced in adjacent slots are somewhat out of phase with each other, because the crest of the flux reaches different slots at different times. Therefore, the resultant voltage of the machine is somewhat smaller than that according to the preceding formula. The influence of the distribution of the winding in the slots is taken into account by multiplying the value of E in the preceding formula by a coefficient k_b, which is smaller than unity and which is called the *breadth factor*. Introducing this factor, and assuming $\chi = 1.11$, which is accurate enough for good commercial alternators, we obtain

$$E = 4.44k_bfN\Phi 10^{-2}, \quad . \quad . \quad . \quad . \quad (31)$$

where Φ is now in megalines. Values of k_b are given in the articles that follow.

Formula (31) applies equally well to the polyphase induction motor or generator. There we also have a uniformly revolving flux in the air-gap, the flux density being distributed in space, according to the sine law. This gliding flux induces e.m.fs. in the stator and rotor windings. The only difference between the two kinds of machines is that in the synchronous alternator the field is made to revolve by mechanical means, while in an induction machine the field is excited by the polyphase currents flowing in

[1] For the proportions of a pole-shoe which very nearly give a sine wave see Arnold, *Wechselstromtechnik*, Vol. 3, p. 247.

the stator and rotor windings. The formulæ of this chapter also apply without change to the synchronous motor, because the construction and the operation of the latter are identical with those of an alternator; the only difference being that an alternator transforms mechanical energy into electrical energy, while a synchronous motor transforms energy in the reverse direction. In all cases the induced voltage is understood and not the line voltage. The latter may differ considerably from the former, due to the impedance drop in the stator winding.

Prob. 15. A delta-connected, 2300 v., 60-cycle, 128.5-r.p.m. alternator is estimated to have a useful flux of about 3.9 megalines per pole. If the machine has one slot per pole per phase how many conductors per slot are needed? Ans. 8.

Prob. 16. A 100,000 cycle alternator for wireless work has one conductor per pole and 600 poles. The rated voltage at no load is 110 v. What is the flux per pole and the speed of the machine?
Ans. 82.5 maxwells; 20,000 r.p.m.

Prob. 17. It is desired to design a line of induction motors for a peripheral speed of 50 met. per sec., the maximum density in the air-gap to be about 6 kilolines per sq.cm. What will be the maximum voltage induced per meter of active length of the stator conductors? Hint: Use formula (27). Ans. 30 volt.

Prob. 18. Formula (31) is deduced under the assumption that each armature conductor is subjected to the " cutting " action of the whole flux. In reality, almost the whole flux passes through the teeth between the conductors, so that it may seem upon a superficial inspection that little voltage could be induced in the conductors which are embedded in slots. Show that such is not the case, and that the same average voltage is induced in the conductors placed in completely closed slots, as in the conductors placed on the surface of a smooth-body armature. Hint: When the flux moves, the same amount of magnetic disturbance must pass in the tangential direction through the slots as through the teeth.

Prob. 19. Deduce eq. (31) directly from eq. (27). Can eq. (31) be derived considering the e.m.f. to be induced by the transformer action?

27. The Breadth Factor. Armature conductors are usually placed in more than one slot per pole per phase, for the following reasons:

(a) The distribution of the magnetic field is more uniform, there being less bunching of the flux under the teeth;

(b) The induced e.m.f. has a better wave form;

(c) The leakage reactance of the winding is reduced;

(d) The same armature punching can be used for machines with different numbers of poles and phases;

(e) The mechanical arrangement and cooling of the coils is somewhat simplified.

The disadvantage of a large number of slots is that more space is taken up by insulation, and the machine becomes more expensive, especially if it is wound for a high voltage. The electromotive force is also somewhat reduced because the voltages induced in different slots are somewhat out of phase with one another. The advantages of a distributed winding generally outweigh its disadvantages, and such windings are used almost entirely. Thus, it is of importance to know how to calculate the value of the breadth factor k_b for a given winding.

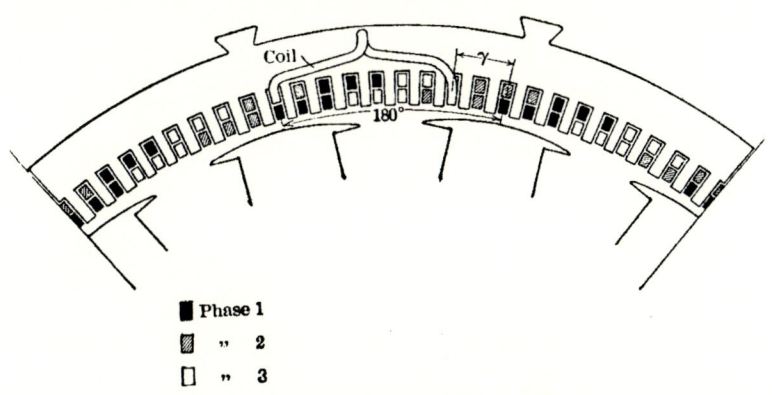

Fig. 16.—A fractional-pitch winding.

In the winding shown in Fig. 15 each conductor is connected with another conductor situated at a distance exactly equal to the pole pitch. It is possible, however, to connect one armature conductor to another at a distance somewhat smaller than the pole pitch (Fig. 16). Such a winding is called a *fractional-pitch winding*, in distinction to the winding shown in Fig. 15; the latter winding is called a full-pitch or hundred-per cent pitch winding. It will be seen from Fig. 16 that, with a two-layer fractional-pitch winding, some slots are occupied by coils belonging to two different phases. The advantages of the fractional-pitch winding are:

(a) The end-connections of the winding are shortened, so that there is some saving in armature copper.

(b) The end-connections occupy less space in the axial direction of the machine, so that the whole machine is shorter.

(c) In a two-pole or four-pole machine it is necessary to use a fractional-pitch winding in order to be able to place machine-wound coils into the slots.

A disadvantage of the fractional-pitch winding is that the e.m.fs. induced on both sides of the same coils are not exactly in phase with each other, so that for a given voltage a larger number of turns or a larger flux is required than with a full-pitch winding. Fractional-pitch windings are used to a considerable extent both in direct- and in alternating-current machinery.

Thus, the induced e.m.f. in an alternator or an induction motor is reduced by the distribution of the winding in more than one slot, and also by the use of a fractional-winding pitch. It is therefore convenient to consider the breadth factor k_b as being equal to the product of two factors, one taking into account the number of slots, and the other the influence of the winding pitch. We thus put

$$k_b = k_s k_w, \qquad \ldots \ldots \ldots \quad (32)$$

where k_s is called the *slot factor* and k_w the *winding-pitch factor*.

For a full-pitch winding $k_w = 1$, and $k_b = k_s$; for a fractional-pitch unislot winding $k_s = 1$, and $k_b = k_w$. The factors k_s and k_w are independent of one another, and their values are calculated in the next two articles.

28. The Slot Factor k_s. Let the stator of an alternator (or induction motor) have two slots per pole per phase, and let the centers of the adjacent slots be displaced by an angle α, in electrical degrees, the pole pitch, τ, corresponding to 180 electrical degrees. If E (Fig. 17) is the vector of the effective voltage induced in the conductors in one slot, the voltage E' due to the conductors in both slots is represented graphically as the geometric sum of two vectors E relatively displaced by the angle α. We see from the figure that $\frac{1}{2}E' = E \cos \frac{1}{2}\alpha$, or $E' = 2E \cos \frac{1}{2}\alpha$. If both sets of conductors were bunched in the same slot we would then have $E' = 2E$. Hence, in this case the coefficient of reduction in voltage, or the slot factor, $k_s = \cos\frac{1}{2}\alpha$.

Let now the armature stamping have S slots per pole per phase, the angle between adjacent slots being again equal to α electrical degrees. Let the vectors marked E in Fig. 18 be the voltages induced in each slot; the resultant voltage E' is found as the geometric sum of the E's. The radius of the circle $r = \frac{1}{2}E/\sin\frac{1}{2}\alpha$, and $\frac{1}{2}E' = r\sin\frac{1}{2}S\alpha$. Therefore,

$$k_s = E'/SE = (\sin \tfrac{1}{2}S\alpha)/(S \sin \tfrac{1}{2}\alpha). \quad . \quad . \quad . \quad (33)$$

When $S = 2$, the formula (33) becomes identical with the expression given before.

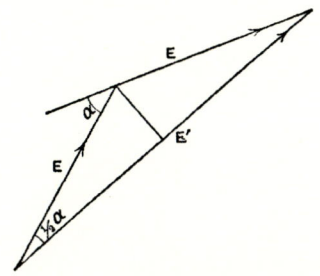

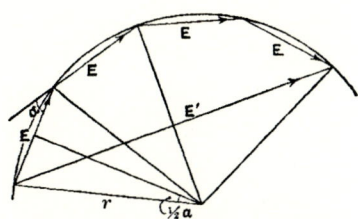

FIG. 17.—A diagram illustrating the slot factor with two slots.   FIG. 18.—A diagram illustrating the slot factor with several slots.

The angle α depends upon the number of slots and the number of phases. Let there be m phases; then $\alpha Sm = 180$ degrees, and

$$\alpha = 180°/(Sm). \quad . \quad . \quad . \quad . \quad (34)$$

The values of k_s in the table below are calculated by using the formulæ (33) and (34).

VALUES OF THE SLOT FACTOR k_s

Slots per Phase per Pole.	Single-phase Winding.	Two-phase Winding.	Three-phase Winding.
1	1.000	1.000	1.000
2	0.707	0.924	0.966
3	0.667	0.911	0.960
4	0.653	0.907	0.958
5	0.647	0.904	0.957
6	0.643	0.903	0.956
Infinity	0.637	0.900	0.955

In single-phase alternators part of the slots are often left empty so as to reduce the breadth of the winding and therefore increase the value of k_s. For instance, if a punching is used with six slots per pole, perhaps only three or four adjacent slots are occupied. In this case, it would be wrong to take the values of k_s from the first column of the table. If, for instance, three slots out of six are occupied, the value of k_s is the same as for a two-phase winding with three slots per pole per phase.

Prob. 20. Check some of the values of k_s given in the table above.

Prob. 21. The armature core of a single-phase alternator is built of stampings having three slots per pole; two slots per pole are utilized. What is the value of k_s?   Ans. 0.866.

Prob. 22. A single-phase machine has S uniformly distributed slots per pole, of which only S' are used for the winding. What is the value of k_s? Ans. Use S' in eq. (33) instead of S; preserve S in eq. (34).

Prob. 23. A six-pole, 6600 v., Y-connected, 50-cycle turbo-alternator is to be built, using an armature with 90 slots. The estimated flux per pole is about 6 megalines. How many conductors are required per slot?   Ans. 20.

Prob. 24. What is the value of k_s when the winding is distributed uniformly on the surface of a smooth-body armature, each phase covering β electrical degrees? Solution: Referring to Fig. 18, k_s is in this case equal to the ratio of the chord E' to the arc of the circle which it subtends. The central angle is β, and we have $k_s = (\sin\frac{1}{2}\beta)/(\frac{1}{2}\beta\pi/180°)$. In a three-phase machine $\beta = 60$ degrees, and therefore $k_s = 0.955$. This is the value given in the last column of the table above.

Prob. 25. Deduce the expression for k_s given in the preceding problem directly from formula (33). Solution: Substituting $S\alpha = \beta$; $S = \infty$ and $\alpha = 0$, an indeterminate expression, $0.\infty$, is obtained. But when the angle α approaches zero its sine is nearly equal to the arc, so that the denominator of the right-hand side of eq. (33) approaches the value $S.\frac{1}{2}\alpha = \frac{1}{2}\beta$, where β is in radians. Changing β to degrees, the required formula is obtained.

29. The Winding-pitch Factor k_w.

Let the distance between the two opposite sides of a coil (Fig. 16) be $180 - \gamma$ degrees, where γ is the angle by which the winding-pitch is shortened. The voltages induced in the two sides of the coil are out of phase with each other by the angle γ, so that if the voltage induced in each side is e, the total voltage is equal to $2e\cos\frac{1}{2}\gamma$ (Fig. 17). Fig. 17 will apply to this case if we read γ for the angle α. Hence, we have that

$$k_w = \cos\frac{1}{2}\gamma. \qquad \ldots \ldots \ldots (35)$$

In practice, the winding-pitch is measured in per cent, or as a fraction of the pole pitch τ. For instance, if there are nine slots per pole and the coil lies in slots 1 and 8, the winding-pitch is 7/9, or 77.8 per cent. If the coil were placed in slots 1 and 10 we would have a full-pitch, or a 100 per cent pitch winding. Let in general the winding-pitch be ζ, expressed as a fraction. Then $\gamma = (1 - \zeta)180°$. Substituting this value of γ into formula (35) we obtain

$$k_w = \cos\,[90°(1 - \zeta)] \quad . \quad . \quad . \quad . \quad . \quad (36)$$

The values of k_w given in Fig. 19 have been calculated according to this formula.

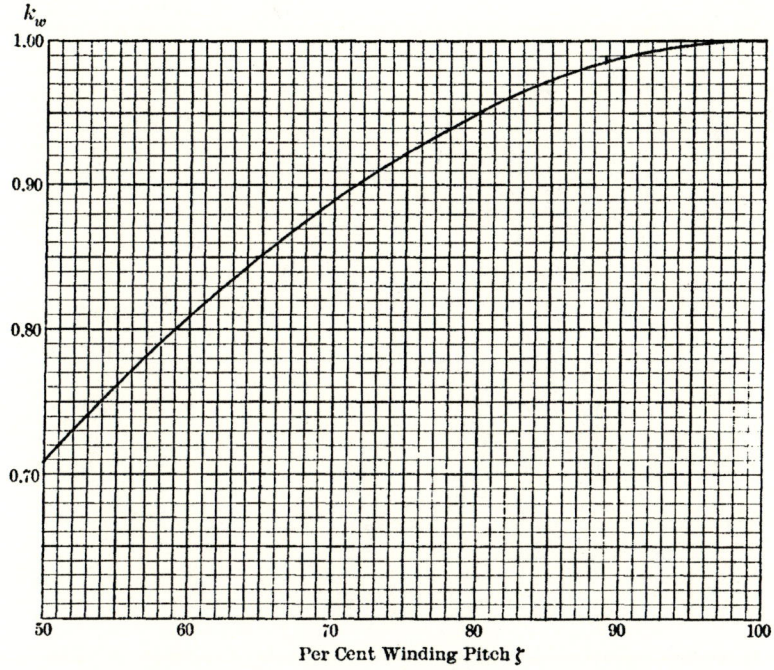

FIG. 19.—Values of the winding-pitch factor k_w.

In applications, one takes the value of k_s from the table, assuming the winding pitch to be one hundred per cent, and multiplies it by the value of k_w taken from the curve (Fig. 19). With fractional-pitch two-layer windings the value of k_s corresponds to the number of slots *per layer* per pole per phase, and not to the total number of slots per pole per phase. This is clear from the explanation given in the preceding paragraph. Thus, for instance, in Fig. 16, k_s must be taken for three slots and not for five slots. If one has to calculate the values of k_b often, it is advisable to plot a set of curves, like the one in Fig. 19, each curve giving the values of k_b for a certain number of slots per pole per phase, against per cent winding pitch as abscissae.

Prob. 26. In a 4-pole, 72-slot, turbo-alternator the coils lie in slots 1 and 13. What is the per cent winding-pitch and by what percentage is the e.m.f. reduced by making the pitch short instead of 100%?
Ans. 66.7 per cent; $1 - k_w = 13.4$ per cent.

Prob. 27. What is the flux per pole at no load in a 6600-volt, 25-cycle, 500 r.p.m., Y-connected induction motor which has 90 slots, 36 conductors per slot, and a winding-pitch of about 73 per cent?
Ans. 7.26 megalines.

Prob. 28. Show that for a chain winding k_w is always equal to unity, in spite of the fact that some of the coils are narrower than the pole pitch.

Prob. 29. Draw a sketch of a single-layer, fractional-pitch winding, using alternate slots for the overlapping phases. Show what values of k_s and k_w should be used for such a winding.

30. Non-sinusoidal Voltages.

In the foregoing calculations the supposition is made that the flux density in the air-gap is distributed according to the sine law so that sinusoidal voltages are induced in each conductor. Under these circumstances the resultant voltage also follows the sine law, no matter what the winding-pitch and the number of slots are. The flux is practically sinusoidal in induction motors because the higher harmonics of the flux are wiped out by the secondary currents induced in the low-resistance rotor. But in synchronous alternators and motors with projecting poles the distribution of the flux in the air-gap is usually different from a pure sine wave. For instance, when the pole shoe is shaped by a cylindrical surface concentric with that of the armature, the air-gap length and consequently the flux density are constant over the larger portion of the pole; therefore, the curve of the field distribution is a flat one. This shape is improved to some extent by chamfering the pole-tips or by shaping the pole shoes to a circle of a smaller radius, so that the length of the air-gap increases gradually toward the pole-tips.

When a machine revolves at a uniform speed, the e.m.f. induced in a single armature conductor has exactly the shape of the field-distribution curve, because in this case the rate of cutting the flux is proportional to the flux density (see eq. 27 above). Therefore, when a machine has but one slot per pole per phase (which condition is undesirable, but unavoidable in low-speed alternators, or in those designed for extremely high frequencies), the shape of the pole-pieces must be worked out very carefully in order to have an e.m.f. approaching the true sine wave. With a larger number of slots this is not so necessary because the. em.fs. induced in different slots are added out of phase with each other, and the undesirable higher harmonics partly cancel each other. The voltage wave is further improved by a judicious use of a fractional-pitch winding. These facts are made clearer in the solution of the problems that follow.[1]

Prob. 30. The flux density in the air-gap under the poles of an alternator is constant for 50 per cent of the pole pitch, and then it drops to zero, according to the straight-line law, on each side in a space of 15 per cent of the pole pitch. Draw to scale the curves of induced e.m.f. for the following windings: (a) Single-phase, one slot per pole; (b) Single-phase, nine slots per pole, five slots being occupied by a one-hundred per cent pitch winding; (c) The same as in (b) only the winding-pitch is equal to 7/9; (d) Three-phase, Y-connected full-pitch winding, two slots per pole per phase; in the latter case give curves of both the phase voltage and the line voltage. On all the curves indicate roughly the equivalent sine wave, in order to see the influence of the number of slots and of the fractional pitch in improving the wave form.

Prob. 31. A three-phase, Y-connected alternator has three slots per pole per phase, and a full-pitch winding. The field curve has an 8 per cent fifth harmonic, that is to say, the amplitude of the fifth harmonic is 0.08 of that of the fundamental sine wave. What is the magnitude of the fifth harmonic in the phase voltage and in the line voltage. Solution: In formula (33) the angle α between the adjacent slots is 20 electrical degrees for the fundamental wave. For the fifth harmonic the same distance between the slots corresponds to 100 electrical degrees. Hence, for the fundamental wave

$$k_s = \sin 30°/(3 \sin 10°) = 0.96;$$

while for the fifth harmonic

$$k_{s5} = \sin 150°/(3 \sin 50°) = 0.217.$$

This means that, due to the distribution in three slots, the fundamental wave of the voltage is reduced to 0.96 of its value in a unislot machine, while the fifth harmonic is reduced to only 0.217 of its corresponding value. Therefore, the relative magnitude of the fifth harmonic in the phase voltage is $8 \times 21.7/96 = 1.8$ per cent, which means that the fifth harmonic is reduced to less than one-fourth of its value in the field curve. In calculating the line voltage the vectors of the fundamental waves in a three-phase machine are combined at an angle of 120 degrees. Conse-

---

[1] For further details see Professor C. A. Adams' paper on " Electromotive Force Wave-shape in Alternators," *Trans. Amer. Inst. Elec. Engs.*, Vol. 28 (1909), Part II, p. 1053.

quently, the vectors of the fifth harmonic are combined at an angle of $120 \times 5 = 600$ degrees, or what is the same, -120 degrees. Therefore the proportion of the fifth harmonic in the line voltage is the same as that in the phase voltage.

Prob. 32. Solve the foregoing problem when the winding pitch is 7/9. Ans. 0.33 per cent. This shows that by properly selecting the winding pitch an objectionable higher harmonic can be reduced to a negligible amount.

Prob. 33. Show that the line voltage of a Y-connected machine can have no 3d, 9th, 15th, etc. harmonics, that is to say, harmonics the numbers of which are multiples of 3, no matter to what extent such harmonics are present in the induced e.m.fs. in each phase.

Prob. 34. Prove that in order to have even harmonics in the induced e.m.f. of an alternator two conditions are necessary: (a) the flux distribution under the alternate poles must be different; (b) the distribution of the armature conductors under the alternate poles must also be different from one another. Indicate pole shapes and an arrangement of the armature winding particularly favorable for the production of the second harmonic. *Note:* In spite of a different distribution of flux densities the total flux is the same under all the poles. Therefore, the average voltages for both half cycles are equal (see Art. 24), though the shape of the two halves of the curve may be different, due to the presence of even harmonics. This shows that there is no " continuous-voltage component " in the wave, or rather that the voltage is in no sense unidirectional, and that a direct-current machine cannot be built with alternate poles without the use of some kind of a commutating device.

31. The Induced E.M.F. in a Direct-current Machine. The

e.m.f. induced in the armature coils of a direct-current machine (Fig. 20) is alternating, but due to the commutator, the voltage between the brushes of opposite polarity remains constant. This voltage is equal at any instant to the sum of the instantaneous e.m.fs. induced in the coils which are connected in series between the brushes. When a coil is transferred from one circuit to another, a new coil in the same electromagnetic position is introduced into the first circuit, and in this wise the voltage between the brushes is maintained practically constant, except for the small variations which occur while the armature is coming back to a symmetrical position. These variations are due to the coils short-circuited by the brushes and to the fact that the number of commutator segments is finite.

Thus, to obtain the value of the voltage between the brushes, it is necessary to find the sum of the e.m.fs. induced at some instant in the individual armature coils which are connected in

series between the brushes.   Each e.m.f. represents an instantaneous value of an alternating e.m.f.; the e.m.fs. induced in different
coils differing in phase from one another, because they occupy
different positions with respect to the poles.   The voltages induced
in the extreme coils of an armature circuit differ from one another
by one-half of a cycle.

Instead of adding the actual instantaneous voltages, it is sufficient to calculate the average voltage per coil, and to multiply it
by the number of coils in series, because the wave form of the
e.m.fs. induced in all the coils is the same, and their phase differ-

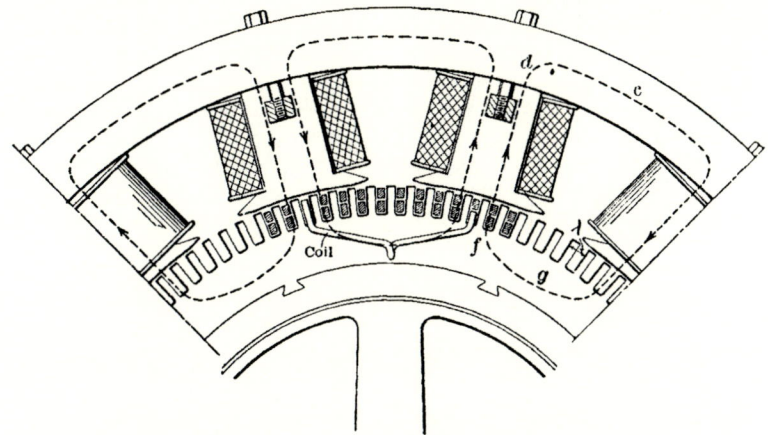

FIG. 20.—The cross-section of a direct-current machine.

ence is distributed uniformly over one-half of a cycle.   According
to eq. (26) the average voltage per turn per half a cycle is $2\Phi/\frac{1}{2}T$,
where $\frac{1}{2}T$ is the time during which the coil moves by one pole
pitch, and Φ is the flux per pole, in webers.   Substituting $1/f$ for
T, the average voltage per turn is equal to $4f\Phi$.   Let there be N
turns in series between the brushes of opposite polarity; then the
induced voltage of the machine is

$$E = 4fN\Phi \times 10^{-2}, \quad \ldots \ldots \quad (37)$$

where Φ is now in megalines.   Thus, in a direct-current machine
the induced voltage between the brushes depends only upon the
total useful flux per pole, and not upon its distribution in the air-
gap.

The relation between the number of turns in series and the total number of turns on the armature depends upon the kind of the armature winding.[1] If the armature has a multiple winding N is equal to the total number of turns on the armature divided by the number of poles. For a two-circuit winding the number of turns in series is equal to one-half of the total number of turns. The number of poles and the speed of the machine do not enter explicitly into formula (37), but are contained in the value of f.

Prob. 35. A 90-slot armature is to be used for a 6-pole, 580-r.p.m., 250-v., direct-current machine with a multiple winding. How many conductors per slot are necessary if the permissible flux per pole is about 3 megalines?   Ans. 10.

Prob. 36. A 550-v., 4-pole railway motor has a two-circuit armature winding which consists of 59 coils, 8 turns per coil. The total resistance of the motor is 0.235 ohm. When the motor runs at 675 r.p.m. it takes in 81 amp. What is the flux per pole at this load?   Ans. 2.5 ml.

Prob. 37. Show that in a direct-current machine the use of a fractional-pitch winding has no effect whatever upon the value of the induced e.m.f., as long as the winding-pitch somewhat exceeds the width of the pole shoe.

Prob. 38. Prove that formula (37) is identical with the expression

$$E = (p/p')(\text{r.p.m.}/60)C\Phi \times 10^{-2}. \quad \ldots \quad \ldots \quad (38)$$

where C is the total number of armature conductors, p is the number of poles, and p' is the number of circuits in parallel.

Prob. 39. Show that the induced e.m.f. is the same when the armature conductors are placed in open or in closed slots as when they are on the surface of a smooth-body armature. See Prob. 18, Art. 26.

Prob. 40. Considerable effort has been made to produce a direct-current generator with alternate poles, and without any commutator. One of the proposals which is sometimes urged by a beginner is to use an ordinary alternator, and to supply the exciting winding with an alternating current of the synchronous frequency. The apparent reasoning is that the field being reversed at the completion of one alternation the next half wave of the induced e.m.f. must be in the same direction as the preceding one, thus giving a unidirectional voltage. Show that such a machine in reality would give an ordinary alternating voltage of double the frequency. Hint: Make use of the fact that an alternating field can be replaced by two constant fields revolving in opposite directions. Or else give a rigid mathematical proof by considering the actual rate at which the armature conductors are cut by the field, which field is at the same time pulsating and revolving.

---

[1] For details concerning the direct-current armature windings see the author's *Experimental Electrical Engineering*, Vol. 2, Chapter 30.

Prob. 41. Prove that if in the preceding problem the frequency of the rotation of the poles is f_1, and the frequency of the alternating current in the exciting winding is f_2, that the voltage induced in the armature is a combination of two waves having frequencies of $f_1 + f_2$ and $f_1 - f_2$ respectively.

32. The Ratio of A.C. to D.C. Voltage in a Rotary Converter.

A rotary converter resembles in its general construction a direct-current machine, except that the armature winding is connected not only to the commutator, but also to two or more slip rings.[1] When such a machine is driven mechanically it can supply a direct current through its commutator, and at the same time an alternating current through its slip rings. It is then called a double-current generator. But if the same machine is connected to a source of alternating voltage and brought up to synchronous speed it runs as a synchronous motor and can supply direct current through its commutator. It is then called a rotary converter. It is also sometimes used for converting direct current into alternating current, and is then called an inverted rotary.

Both the direct and the alternating voltages are induced in the armature of a rotary converter by the same field, and our problem is to find the ratio between the two voltages for a given arrangement of the slip rings. Consider first the simplest case of a single-phase converter with two collector rings connected to the armature winding, at some two points 180 electrical degrees apart. If the armature has a multiple winding each slip ring is connected to the armature in as many places as there are pairs of poles. In the case of a two-circuit winding each collector ring is connected to the armature in one place only.

If the machine has p poles then p times during each revolution the direct-current brushes make a connection with the same armature conductors to which the slip rings are connected. At these moments the alternating voltage is a maximum, because the direct-current brushes are placed in the position where the induced voltage in the armature is a maximum. Thus, with two slip rings, connected 180 electrical degrees apart, the maximum value of the alternating voltage is equal to the voltage on the direct-current side. If the pole shoes are shaped so that the alternating voltage is approximately sinusoidal, the effective value of the voltage between the slip rings is $1/\sqrt{2} = 70.7$ per cent of that between the

[1] See the author's *Experimental Electrical Engineering*, Vol. 2, Chapter 28.

direct-current brushes. The same ratio holds true for a two-phase rotary, for the voltages induced in each phase.

Let now two slip rings be connected at two points of the armature winding, α electrical degrees apart. In order to obtain the value of the alternating voltage the vectors of the voltages induced in the individual coils must be added geometrically, as in Fig. 18. With a large number of coils the chords can be replaced by the arc, and in this way Fig. 21 is obtained.

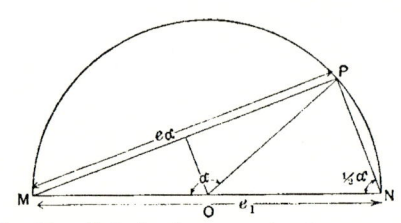

Fig. 21.—Relation between the alternating voltages in a rotary converter.

The diameter $MN = e_1$ of the semicircle represents the vector of the alternating voltage when the points of connection to the slip rings are displaced by 180 electrical degrees, while the chord $MP = e_\alpha$ gives the voltage between two slip rings when the taps are distant by α electrical degrees. It will thus be seen that

$$e_\alpha = e_1 \sin \tfrac{1}{2}\alpha. \qquad \ldots \qquad (39)$$

But we have seen before that $e_1 = 0.707E$, where E is the voltage on the direct-current side of the machine. Hence, for sinusoidal voltages,

$$e_\alpha = 0.707E \sin \tfrac{1}{2}\alpha. \qquad \ldots \qquad (40)$$

The following table has been calculated, using this formula.

Number of slip rings	2	3	4	5	
Angle between the adjacent taps in electrical degrees	180	120	90	72	60
Ratio of alternating to continuous voltage, in per cent	70.7	61.2	50	41.5	35.3

The foregoing theory shows that the ratio of the continuous to the alternating voltage is fixed in a given converter, and in order to raise the value of the direct voltage it is necessary to raise the applied alternating voltage. This is done in practice either by means of various voltage regulators separate from the converter, or by means of a booster built as a part of the converter. Another method of varying the voltage is by using the so-called split-pole converter. In this machine the distribution of the flux density in the air-gap can be varied within wide limits, and consequently that component of the field which is sinusoidal can be varied. The result is that the ratio of the direct to the alternating voltage is also variable. Namely, we have seen before that the value of the continuous voltage does not depend upon the distribution of the flux, but only upon its total value, while the effective value of the alternating voltage depends upon the sine wave or fundamental component of the flux distribution.[1]

Prob. 42. Check some of the values given in the table above.

Prob. 43. A three-phase rotary converter must deliver direct current at 550 v. What is the voltage on the alternating-current side?

Ans. 337 v.

Prob. 44. The same rotary is to be tapped in three additional places so as to get two-phase current also. How many different voltages are on the alternating-current side and what are they?

Ans. 389, 337, 275, 195, 100.

Prob. 45. The table given above holds true only when the flux density is distributed approximately according to the sine law. Show how to determine the ratio of alternating to continuous voltage in the case of two collector rings connected to taps 180 electrical degrees apart, when the curve of field distribution is given graphically. Solution: Divide the pole pitch into a sufficient number of equal parts and mark them on a strip of paper. Place the strip along the axis of abscissæ. The sum of the ordinates of the flux-density curve, corresponding to the points of division, at a certain position of the strip, gives the instantaneous value of the alternating voltage. Having performed the summation for a sufficient number of positions of the strip, the wave of the induced e.m.f. is plotted. The scale of the curve is determined by the condition that the maximum ordinate is equal to the value of the continuous voltage. The effective value is found in the well-known way, either in rectangular or in polar coordinates (see the *Electric Circuit*, Arts. 17 and 18).

Prob. 46. Apply the solution of the preceding problem to the field distribution specified in Prob. 30, Art. 30.   Ans. 81.5 per cent.

Prob. 47. Extend the method described in Prob. 45 to the case when the distance between the taps connected to the slip rings is less than 180 degrees. Show how to find the scale of voltage.

Prob. 48. How does a fractional pitch affect the values given in the table above, and the solution outlined in Prob. 45?

Prob. 49. Show how to solve problems 45 to 48 when the field curve is given analytically, as $B = F(\alpha)$, for instance in the form of a Fourier series. Hint: See C. A. Adams, " Voltage Ratio in Synchronous Converters with Special Reference to the Split-pole Converter," *Trans. Amer. Inst. Elec. Engs.*, Vol. 27 (1908), part II, p. 959.

[1] See C. W. Stone, " Some Developments in Synchronous Converters," *Trans. Amer. Inst. Elec. Engs.*, Vol. 27 (1908), p. 181.

CHAPTER V.

THE EXCITING AMPERE-TURNS IN ELECTRICAL MACHINERY

33. The Exciting Current in a Transformer. The magnetic flux in the core of a constant-potential transformer is determined essentially by the primary applied voltage, and is practically independent of the load (see Art. 25). When the terminal voltage is given, the flux becomes definite as well. The ampere-turns necessary for producing the flux are called the magnetizing or the exciting ampere-turns. When the secondary circuit is open the only current which flows through the primary winding is that necessary for producing the flux. This current is called the *no-load, exciting,* or *magnetizing current* of the transformer. When the transformer is loaded, the vector difference between the primary and the secondary ampere-turns is practically equal to the exciting ampere-turns at no-load.

The exciting current is partly reactive, being due to the periodic transfer of energy between the electric and the magnetic circuits (see Art. 16 above), partly it represents a loss of energy due to hysteresis and eddy currents in the core. Some writers call the reactive component of the no-load current the magnetizing current, and the total no-load current the exciting current. Generally, however, the words magnetizing and exciting are used interchangeably to denote the total no-load current. The components of the current in phase and in quadrature with the induced voltage are called the energy and the reactive components respectively.

The no-load or exciting current in a transformer must usually not exceed a specified percentage of the rated full-load current; it is therefore of importance to know how to calculate the exciting current from the given dimensions of a transformer. Knowing the applied voltage and the number of turns, the maximum value of the flux is calculated from eq. (28). We shall assume first that the maximum flux density in the core is so low, that it lies practically on the straight part of the magnetization curve of the material (Fig. 3). The case of high flux densities is considered in the next article.

Since by assumption the instantaneous magnetomotive forces are proportional to the corresponding flux densities, the magnetizing current must vary according to the sine law. It is sufficient, therefore, to calculate the maximum value of the magnetomotive force, corresponding to the maximum flux. Knowing the amplitude Φ_m of the flux and the net cross-section of the core, A, the flux density B_m becomes known; from the magnetization curve of the material (Fig. 3) the corresponding value of H_m, or the ampere-turns per unit length of path, is found. The mean length l of the lines of force is determined from the drawing of the core, so that the total magnetizing ampere-turns $M_m = H_m l$ can be calculated. The mean magnetic path around the corners is somewhat shorter than the mean geometric path.

Let n_1 be the number of turns in the primary winding, and i_o the *effective* value of the reactive component of the exciting current. We have then

$$i_0 n_1 \sqrt{2} = M_m. \quad . \quad . \quad . \quad . \quad . \quad . \quad (41)$$

From this equation the quantity which is unknown can be calculated.

It is presupposed in the above deduction that the joints between the laminations offer no reluctance. In reality, the contact reluctance is appreciable; its value depends upon the character of the joints, and the care exercised in the assembling of the core. This reluctance of the joints can be expressed by the length of an equivalent air-gap having the same reluctance. Thus, experiments show that each overlapping joint is equivalent to an air-gap 0.04 mm. long. A butt joint, with very careful workmanship, is equivalent to an air-gap of about 0.05 mm.; in practice, a butt joint may offer a reluctance of from 50 to 100 per cent higher than the foregoing value.[1] Knowing the

[1] H. Bohle, "Magnetic Reluctance of Joints in Transforming Iron," *Journal* (British) *Inst. Electr. Engs.,* Vol. 41, 1908, p. 527. It is convenient to estimate the influence of the joints in ampere-turns at a standard flux density. For each lap joint 32 ampere-turns must be added at a density of 10 kilolines per square centimeter, while a butt joint requires at the same

length a of the equivalent air-gap, the number of additional ampere-turns is calculated according to the formula aB_m/μ, and then is multiplied by the number of joints in series (usually four). This number of ampere-turns must be added to M_m calculated above.

The energy component i_1 of the exciting current is determined from the power lost in hysteresis and eddy currents in the core. Having calculated this power P as is explained in Article 19, we find $i_1 = P/E_1$, where E_1 is the primary applied voltage. Knowing i_o and i_1, the total no-load current is found as their geometric sum, $i_{ex}^2 = (i_0^2 + i_1^2)$.

The watts expended in core loss depend only upon the volume of the iron, the frequency, and the flux density used. It can be also shown that the reactive volt-amperes required for the excitation of the magnetic circuit of a transformer depend only upon the volume of the iron, the frequency, and the flux density. Namely, neglecting the influence of the joints, eq. (41) can be written in the form

$$i_0 n_1 \sqrt{2} = H_m l.$$

Eq. (28) in Art. 25 can be written as

$$E_1 = 4.44 n_1 f A B_m \times 10^{-5}$$

where A is the cross-section of the iron, and B_m is the maximum flux density, in kilolines per square centimeter. Multiplying these two equations together, term by term, and cancelling n_1 we get, after reduction,

$$E_1 i_o / V = \pi f B_m H_m \times 10^{-5}, \qquad . \quad . \quad . \quad . \quad (42)$$

where $V = Al$ is the volume of the iron, in cubic centimeters. The left-hand side of eq. (42) represents the reactive magnetizing volt-amperes per unit volume of iron; the right-hand side is a function of f and B_m only, because H_m can be expressed through B_m from the magnetization curve of the material.

Formula (42) can be plotted as a set of curves, one for each commercial frequency. These curves are quite convenient in the design of transformers, because they enable one to estimate directly either the permissible volume of iron, or the permissible flux density, when the reactive component of the exciting cur-

density from 60 to 80 ampere-turns. At other flux densities the increase is proportional.

rent is limited to a certain percentage of the full-load current. In practice, such curves are sometimes plotted directly from the results of tests on previously built transformers. These experimental curves are the most secure guide for predicting the exciting current in transformers; formula (42) shows their rational basis.

Prob. 1. Prove that if there were no core loss the exciting current would be purely reactive, that is to say, in a leading phase quadrature with the induced voltage.

Prob. 2. The core of a 22-kv. 25-cycle transformer, like the one shown in Fig. 12, has a gross cross-section of 4500 sq.cm.; the mean path of the lines of force is 420 cm.; the material is silicon steel; the maximum flux is 36 megalines. The expected reluctance of each of the four butt joints is estimated to be equivalent to an 0.08 mm. air-gap. What are the two components of the exciting current, and what is the total no-load current?   Ans. 1.8; 0.4; 1.85.

Prob. 3. In what respects does the calculation of the magnetizing current in a shell-type or cruciform-type transformer differ from that in a core-type transformer?

Prob. 4. Show that for flux densities up to 10 kl./sq.cm. the magnetizing volt-amperes per kilogram of carbon steel at 60 cycles are approximately equal to $7.3(B_m/10)^2$.

Prob. 5. Show that the influence of the joints can be taken into account in formula (42) by adding to the actual volume of the iron the volume of the air-gaps multiplied by the relative permeability of the iron.

Prob. 6. A shell-type 1000-kva., 60-cycle transformer is to have a core made of silicon-steel punchings of a width $w = 17$ cm. (Fig. 13); the average length of the magnetic path in iron is 180 cm.; the reactive component of the no-load current must not exceed 2 per cent of the full-load current. Draw curves of the required height of the core per link, and of the total core loss in per cent of the rated kva., for flux densities up to 10 kl./sq.cm.
          Ans. $B^2h = 5200$; at $B = 9$, $P = 0.51$ per cent.

34. The Exciting Current in a Transformer with a Saturated Core.

In the preceding article the flux density in the core is supposed to be within the range of the straight part of the saturation curves (Fig. 3), so that, when the flux varies according to the sine law, the magnetizing current also follows a sine wave. We shall now consider the case when the flux density rises to a value on or beyond the knee of the magnetization curve. Such high flux densities are used with silicon steel cores, especially at low frequencies. In this case the magnetizing current does not vary according to the sine law, but is a peaked wave, because at the moments when the flux is approaching its maximum, the current is increasing faster than the flux, on account of saturation. The amplitude factor of the current wave, or the ratio of the amplitude to the effective value is no more equal to $\sqrt{2}$, but is larger. Let this ratio be denoted by χ_a. Then eq. (41) becomes

$$i_0 n_1 \chi_a = M_m, \quad \ldots \ldots \quad (43)$$

where i_0 is as before the effective value of the reactive component of the exciting current. The value of χ_a is obtained by actually

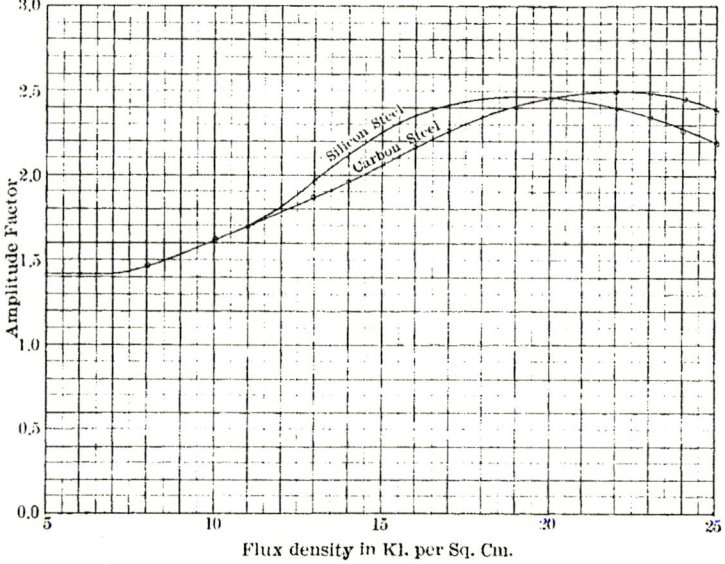

Fig. 22.—Ratio of the amplitude to the effective value of the magnetizing current.

plotting the curve of the magnetizing current from point to point and calculating its effective value. Since the procedure is rather long, it is convenient to calculate the values of χ_a once for all for the working range of values B_m. This has been done for the materials represented in Fig. 3, and the results are plotted in Fig. 22.

Strictly speaking, the exciting current is unsymmetrical, due to the effect of hysteresis, and the values of χ_a ought to be calculated, using the hysteresis loops of the steel. However, it is very nearly correct to calculate χ_a from the magnetization curve, and to calculate the energy component of the exciting current separately, from the core loss curves (Fig. 10). The magnetizing current required for the joints is calculated separately, using $\chi_a = \sqrt{2}$, according to eq. (41). The total *effective* magnetizing current is found by adding together the values of i_0 for the iron and for the joints. The loss component, i_1, is added to this value in quadrature, to get the total no-load current. As is mentioned above, it is preferred in practice to estimate the total exciting current of new transformers from the curves of no-load volt-amperes per kilogram of iron, the values being obtained from tests on similar transformers.

Prob. 7. The core of a 25-cycle cruciform type transformer (Fig. 14) weighs 265 kg.; the mean length of the magnetic path is 170 cm.; the material is silicon steel. The 4400-v. winding of the transformer has 1100 turns in series. What is the reactive component of the no-load current?　　　　　　　　　　　　　　　　Ans. 8.4 amperes.

Prob. 8. Check a few points on the curves in Fig. 22.

Prob. 9. Show that in formula (42) the coefficient π is a special case of the more general factor $4.44/\chi_a$, when the magnetizing current does not follow the sine wave.

Prob. 10. What are the reactive volt-amperes per kilogram of carbon steel at 40 cycles and at a flux density of 16 kl./sq.cm.?　　　Ans. 56.4.

Prob. 11. Show how to calculate the exciting ampere-turns required for a given flux in a thick and short core in which the flux density is different along different paths.

35. The Types of Magnetic Circuit Occurring in Revolving Machinery. The remainder of this chapter and the next chapter have for their object the calculation of the exciting ampere-turns necessary for producing a certain useful flux in the principal types of electric generators and motors. In direct-current machines, in alternators, and in rotary convertors it is necessary to know the exciting or field ampere-turns in order to plot the no-load saturation curve, to predict the performance of the machine under various loads, and to design the field coils. In an induction motor one wants to know the required excitation in order to determine the no-load current, or to calculate the number of turns in the stator winding, when the limiting value of the no-load current is prescribed. The general procedure in determining the required number of ampere-turns for a given flux is in many respects the same in all the types of electrical machinery, so that it is possible to outline the general method before going into details.

In direct-current machines and in synchronous generators, motors, and rotary converters, the magnetic flux (Figs. 15 and 20) from a field pole passes into the air-gap and the armature teeth. In the armature core the flux is divided into two halves, each half going to one of the adjacent poles. The magnetic paths are completed through the field frame. Part of the flux passes directly from one pole to the two adjacent poles through the air, without going through the armature. This part of the flux is known as the *leakage flux*. The closed magnetic paths and the field coils of a machine may be thought of as the consecutive links of a closed chain. While in a transformer the chain is open, in generators

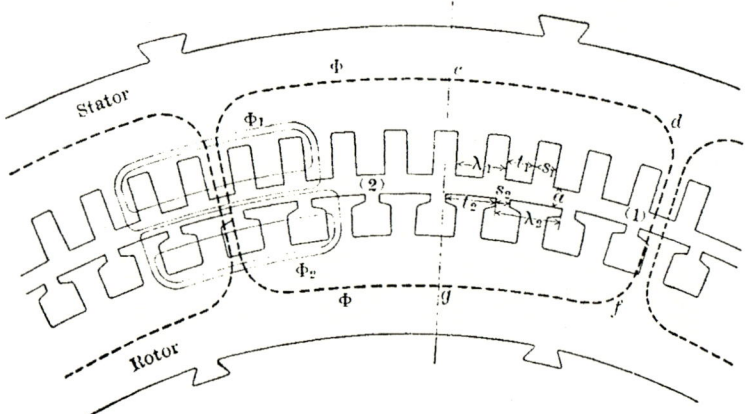

Fig. 23.—The paths of the main flux and of the leakage fluxes in an induction motor (or generator).

and motors the chain must be closed on account of the continuous rotation.

In induction machines, both generators and motors (Fig. 23), the flux at no load is produced by the currents in the stator windings only. When the machine is loaded, the flux is produced by the combined action of the stator and rotor currents, the rotor currents opposing those in the stator, the same as in a transformer. Therefore, the flux in the loaded machine may be regarded as the resultant of the following three component fluxes: The main or useful flux, Φ, which links with both the primary and the secondary windings; the primary leakage flux, Φ_1, which links with the stator winding only; and the secondary leakage flux, Φ_2 which is linked with the rotor winding alone. The leakage fluxes not only do not contribute to the useful torque of the machine, but actually reduce it. In reality, there is of course but one flux, the resultant of the three, but for the purposes of theory and computations the three component fluxes can be considered as if they had a real separate existence. In this and in the following chapter the main flux only will be discussed for this type of machinery. Consideration of the leakage flux will be reserved to Art. 66.

The total magnetomotive force per magnetic circuit is equal to the sum of the m.m.fs. necessary for establishing the required flux in the separate parts of the circuit which are in series, viz., the pole-pieces, the air-gap, the teeth, and the armature core. All the necessary elements for the solution of this problem have been discussed in the first two chapters. It remains here to establish some semi-empirical " short-cut " rules and formulæ for the irregular parts of the circuit, for which, although close approximations can be made, the exact solution is either impossible or too complicated for the purposes of this text. The following topics are considered more in detail in the subsequent articles of this and of the following chapter.

(a) The ampere-turns necessary for the air-gap when it is limited on one side or on both sides by teeth, so that the flux density in the air-gap is not uniform.

(b) The ampere-turns necessary for the armature teeth when they are so highly saturated that an appreciable part of the flux passes through the slots between the teeth.

(c) The ampere-turns necessary for the highly saturated cores in which the lengths of the individual paths differ considerably from one another, with a consequent lack of uniformity in the flux density.

(d) The leakage coefficient and the value of the leakage flux which passes directly from pole to pole. This leakage flux increases the flux density in the poles and in the field frame of the machine, and consequently increases the required number of ampere-turns.

All of the m.m.f. calculations that follow are *per pole* of the machine, or what is the same, for one-half of a complete magnetic circuit (*cdfg* in Figs. 15, 20, and 23), the two halves being identical. This fact must be borne in mind when comparing the formulæ with those given in other books, in which the required ampere-turns are sometimes calculated for a complete magnetic circuit.

Prob. 12. Inspect working drawings of electrical machines found in various books and magazine articles; indicate the paths of the main and of the leakage fluxes; and make clear to yourself the reasons for the use of different kinds of steel and iron in the frame, the core, the pole-pieces, and the pole shoes.

Prob. 13. Make sketches of the magnetic circuit of a turbo-alternator with a distributed field winding, of a homopolar machine, of an inductor-type alternator, and of a single-phase commutator motor. Indicate the paths of the useful and of the leakage fluxes.

36. The Air-gap Ampere Turns. The general character of the distribution of the magnetic flux in the air-gap of a synchronous and of a direct-current machine is shown in Fig. 24, the curvature of the armature being disregarded. The principal features of this flux distribution are as follows:

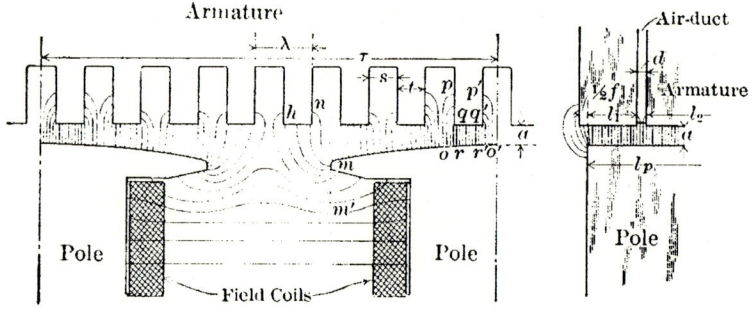

Fig. 24.—The cross-section of a direct-current or synchronous machine, showing the flux in the air-gap.

(*a*) The flux per tooth pitch λ is practically the same under all the teeth in the middle part of the pole, where the air-gap has a constant length, and is smaller for the teeth near the pole-tips where the air-gap is larger.

(*b*) On the armature surface the flux is concentrated mainly at the tooth-tips; very few lines of force enter the armature through the sides and the bottom of the slots.

(*c*) There is a considerable spreading, or fringing, of the lines of force at the pole-tips.

(*d*) In the planes passing through the axis of the shaft of the machine there is also some spreading or fringing of the lines of force at the flank surfaces of the armature and the pole, and in the ventilating ducts.

This picture of the flux distribution follows directly from the fundamental law of the magnetic circuit, the flux density being higher at the places where the permeance of the path is higher. The actual flux distribution is such that the total permeance of all the paths is a maximum, as compared to any other possible distribution. In other other words, the flux distributes itself in such a way, that with a given m.m.f. the total flux is a maximum, or with a given flux the required m.m.f. is a minimum. This is confirmed by the beautiful experiments of Professor Hele-Shaw and his collaborators,[1] who have obtained photographs of the stream lines of a fluid flowing through an arrangement which imitated the shape and the relative permeances of the air-gap and of the teeth in an electric machine.

Let $\mathcal{P}_a$ be the total permeance in perms of the air-gap between the surface of the pole shoe and the teeth, and let Φ be the useful flux per pole, in maxwells, which is supposed to be given. Then, according to eq. (2), Art. 5, the number of ampere-turns required for the air-gap is

$$M_a = \Phi / \mathcal{P}_a. \qquad . \quad . \quad . \quad . \quad . \quad . \quad (44)$$

The problem is to calculate the permeance of the gap from the drawing of the machine.

One of the usual practical methods is to calculate $\mathcal{P}$ under certain simplifying assumptions and then multiply the result by an empirical coefficient determined from tests on similar machines. The simplest assumptions are (Fig. 25): (a) that the armature has a smooth surface, the slots being filled with iron of the same permeability as that of the teeth; (b) that the external surface of the pole shoes is concentric with that of the armature; (c) that the equivalent air-gap a_{eq} is equal to two-thirds of the minimum air-gap plus one-third of the maximum air-gap of the actual machine; (d) that the ventilating ducts are filled with iron; (e) that the paths of the fringing flux at the edges of the pole shoe are straight lines, and extend longitudinally to the edge of the armature surface and laterally for a distance equal to the equivalent air-gap on each side.

[1] For a detailed account of the experimental and theoretical investigations on this subject, with numerous references, see Hawkins and Wallis, *The Dynamo* (1909), Vol. 1, Chapter XV.

With these assumptions, the permeance of the " simplified " air-gap is

$$\mathcal{P}_s = 1.25 w_s l_s / a_{eq}, \qquad \qquad (45)$$

where w_s is the average width of the flux, and l_s is its average axial length. Or

$$w_s = \tfrac{1}{2}(w_a + w_p) = w_p + a';$$

$$l_s = \tfrac{1}{2}(l_a + l_p);$$

$$a_{eq} = \tfrac{2}{3} a_{min} + \tfrac{1}{3} a_{max}.$$

The lateral spread a' of the lines of force at each pole-tip is taken to be approximately equal to a_{eq}.

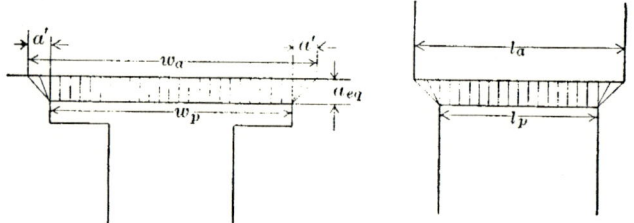

FIG. 25.—Magnetic flux in the simplified air-gap.

The permeance of the actual air-gap is smaller than that of the simplified gap, so that we have

$$\mathcal{P}_a = \mathcal{P}_s / k_a, \qquad \qquad (46)$$

where k_a is a coefficient larger than unity, called the *air-gap factor*. Substituting the value of $\mathcal{P}$ from eq. (46) into (44) gives

$$M_a = k_a \Phi / \mathcal{P}_s \qquad \qquad (47)$$

so that k_a is the factor by which the ampere-turns for the simplified air-gap must be multiplied in order to obtain the ampere-turns required for the actual air-gap. The value of k_a usually varies between 1.1 and 1.3, depending on the relative proportions of the teeth, the slots, and the air-gap, and on the shape of the poles.

The numerical values of k_a are calculated from the results of tests on machines of proportions similar to that being computed. Let the no-load saturation curve of a machine be available from test; this is a curve which gives the relation between the induced voltage and the field current of the machine. From the known specifications of the machine this curve can be easily converted into one which gives the useful flux per pole against the ampere-turns per pole as abscissæ. The lower part of such a curve is always a straight line, there being then practically no saturation in the iron. On this part of the curve, practically the whole m.m.f. is consumed in the air-gap, so that the actual permeance of the air-gap is found by dividing one of the ordinates by the corresponding abscissa. The permeance of the simplified air-gap is calculated from eq. (45), and the ratio of the two gives the value of the coefficient k_a. This value is then used in the design and calculation of the performance of new machines with similar proportions. Engineering judgment and practical experience are factors of considerable importance in estimating the values of k_a for new machines.

The same method of calculating the air-gap ampere-turns is applicable to induction machines (Fig. 23). The ampere-turns are computed, assuming both the rotor and the stator to have smooth iron surfaces, without slots; the result is then multiplied by a factor k_a larger than unity, determined from tests upon machines of similar proportions.

A more accurate, though more elaborate, method for calculating the air-gap ampere-turns is explained in the next article.

Prob. 14. Calculate the air-gap ampere-turns per pole for a 6600 v., 25-cycle, 375-r.p.m. alternator to be built according to the following specifications: The bore 2.4 m.; the gross axial length of the armature core 55 cm.; seven air ducts 9 mm. each; the minimum air gap is 15 mm.; the maximum air gap is 24 mm. The poles cover 66 per cent of the periphery; the axial length of the pole shoes is 53 cm. The useful flux per pole at no-load and at the rated voltage is 19.1 megalines. The air-gap factor is estimated to be about 1.15.   **Ans.** 9100

Prob. 15. The no-load characteristic obtained from the test upon the machine specified in the preceding problem has a straight part such that at a field current of 45 amp. the line voltage is 4000 v. Each field coil has 120 turns. What is the true value of the air-gap factor?
    Ans. 1.125.

Prob. 16. A pole shoe is so shaped that the minimum air-gap is a_0 and the maximum air-gap is $a_1 = a_0 + \Delta a$, the increase in the length being proportional to the square of the distance from the center of the pole. What is the length a_{cq} of the equivalent uniform air-gap such that its total permeance is the same as that of the given air-gap? Assume a smooth-body armature, and neglect the fringing at the pole-tips. Solution: Let the peripheral width of the pole be $2w$; then the length of the air-gap at a distance x from the center is $a_x = a_0 + Ja(x/w)^2$. The permeance of an infinitesimal path of the width dx is proportional to dx/a_x. Hence we have the relation

$$\int_0^w dx/[a_0 + Ja(x/w)^2] = w/a_{eq},$$

from which

$$a_{eq} = \backslash\ a_0 Ja\ \tan^{-1}(\backslash\ Ja\ a_0).$$

Prob. 17. What is the length of the equivalent air-gap in the preceding problem if the clearance at the pole-tips is twice the clearance at the center of the pole?   Ans. $1.273a_0$.

Prob. 18. Show that, when the air-gap is non-uniform, the length of the equivalent uniform gap can be determined approximately, according to Simpson's Rule, from the equation

$$a_{eq}^{-1} = \tfrac{1}{6}(a_0^{-1} + 4a_m^{-1} + a_1^{-1}).$$

where a_0, a_1, and a_m are the lengths of the gap at the center, at the tip of the pole, and midway respectively. If the air-gap is uniform under the major portion of the pole, but the pole shoe is chamfered, more terms must be taken in Simpson's formula in order to obtain a_{eq} with a sufficient accuracy.

Prob. 19. What is the length of the air-gap required in problems 16 and 17, according to the formula given in problem 18?

             Ans. $1.276a_0$.

37. The Method of Equivalent Permeances for the Calculation of Air-gap Ampere-turns.

An inspection of Fig. 24 will show that the total permeance of the air-gap is made up of a number of permeances in parallel. It is equal therefore to the sum of these permeances. For the purpose of calculation two kinds of permeances are considered separately: those from the teeth to the pole surface proper, and those from the teeth to the pole-tips. The former can be calculated quite accurately, the latter are to some extent estimated.

The permeance per tooth pitch in the part of the air-gap near the center of the pole can be divided into two parts, that under the tooth-tip, and the fringe from the sides of the slots and in the ventilating ducts. The permeance of the paths which proceed from the tooth-tip constitutes the larger portion and is made up of nearly parallel lines; this permeance is therefore easily computed. The values of the permeance of the fringe from the flank of the tooth to the perpendicular surface of the pole have been deter-

mined theoretically by Mr. F. W. Carter.[1] Only the numerical
results are given here, in a somewhat simplified practical form;
the solution itself presupposing a knowledge of the properties of
conjugate functions.[2]

Consider the permeance of two tooth fringes, such as $opqr$ and
$o'p'q'r'$ (Fig. 24), perpendicular to the plane of the paper. This
permeance depends only upon the ratio of the slot width s to the
length a of the air-gap, for let both s and a be increased say twice:
The length and the cross-section of each elementary tube of force
is also increased twice, hence its permeance remains the same.

The permeance of each fringe can be replaced by the permeance
of an equivalent rectangular path of the length a and of a width
$\frac{1}{2}\Delta t$ (Fig. 26). This is the same as increasing the width of the
tooth by the amount Δt and assuming all the lines of force to be
parallel to each other in the air-gap. The permeance of the path
which replaces the two fringes is equal to $\mu \Delta t/a$. From what has
been said above follows that the ratio $\Delta t/a$ depends only upon the
ratio of s/a; the relationship between the two ratios is plotted in
Fig. 26, from Carter's calculations. For the sake of convenience
and accuracy, the curve is drawn to two different scales, one for
large the other for small values of s/a.

The curve in Fig. 26 may be interpreted in two ways: It may
be said to represent the "geometric permeance" of the fringe (for
$\mu = 1$); or else it may be said to give the correspondings sets of
values of s and Δt, measured in the lengths of the air-gap as the
unit. With a given a, Δt increases with s, because the maximum
width of the actual fringe is $\frac{1}{2}s$. With a given s the width Δt
increases toward the pole-tip (if the air-gap is variable), because
with a longer air-gap the fringing lines of flux fill a larger part of
the air-gap under the slot.

The corrected width of the tooth is $t' = t + \Delta t$ and the permeance
of the air-gap, in perms per tooth pitch, is

$$\mathcal{P}_{at} = 1.25(\Delta t/a_x + t/a_x)l_{eff}, \quad \cdots \quad (48)$$

[1] Note on Air-gap Induction, *Journ. Inst. Electr. Eng.* (British), Vol. 29,
(1899–1900), p. 929; Air-gap Induction, *Electrical World*, Vol. 38, (1901)
p. 884; See also Hawkins and Wallis, *The Dynamo* (1909), Vol. 1, p. 446;
E. Arnold, *Die Gleichstrommaschine* (1906), Vol. 1, p. 266.

[2] J. C. Maxwell, *Electricity and Magnetism*, Vol. 1, p. 284; J. J. Thomson,
Recent Researches in Electricity and Magnetism, Chapter III; Horace Lamb,
Hydrodynamics (1895), Chapter IV.

where a_x is the length of the air-gap at the center of the tooth, and l_{eff} is the effective axial length of the machine (see below). The value of $\mathit{\Delta}t/a_x$ must be taken from Fig. 26 for the corresponding ratio s/a_x.

The permeance of the pole fringe, hmn in Fig. 24, cannot be calculated by the foregoing method, because this permeance depends upon the irregular shape of the pole-tip. The pole-fringe permeance is usually estimated graphically by drawing lines of force, taking Fig. 24 as a guide[1]; the permeance of each tube of flux between the pole and the armature is $\mu A/l$, where A is the mean cross-section, and l is the mean length of the tube. The fringe permeance is of the order of magnitude of 10 per cent of the total permeance of the air-gap, so that some error in its estimation does not seriously affect the total required ampere-turns. Careful designers sometimes calculate the air-gap permeance for two positions of the pole, differing from each other by one-half of the tooth pitch, and take the average of the two results.

Carter's curve could be used directly for calculating the pole-fringe permeance, if the pole waist were of the same width as the pole shoe (line mm' in Fig. 24), and if the armature had no slots. In this case the space between the adjacent poles could be considered as a big slot, and the curve in Fig. 26 could be directly applied to it. On account of a smaller width of the pole core and because of the armature slots the mean length of the lines of force in the fringe is increased, so that the actual permeance of the pole-fringe is somewhat smaller than that according to Carter's curve. By practice and experience one can acquire a judgment as to what fraction of Carter's permeance to take in a given case.

The length l_{eff} is a sum of the parts such as l_1, l_2, etc. (Fig. 24), on which the lines of force are parallel, and of small additional lengths which take account of the fringing in the air-ducts and at the pole flanks. These additional lengths are again estimated from Carter's curve (Fig. 26). The fringe $\mathit{\Delta}d$ in an air-duct of the width d is practically the same as that in a slot of the width $s = d$. The additional length $\mathit{\Delta}f$ for the pole flanks is found by considering the two fringes as due to a slot of the width f. When the stationary and the revolving parts are of the same axial length so that $f = 0$, there still remains some fringe permeance between the

[1] See Art. 41 below in regard to the drawing of the lines of force by the judgment of the eye.

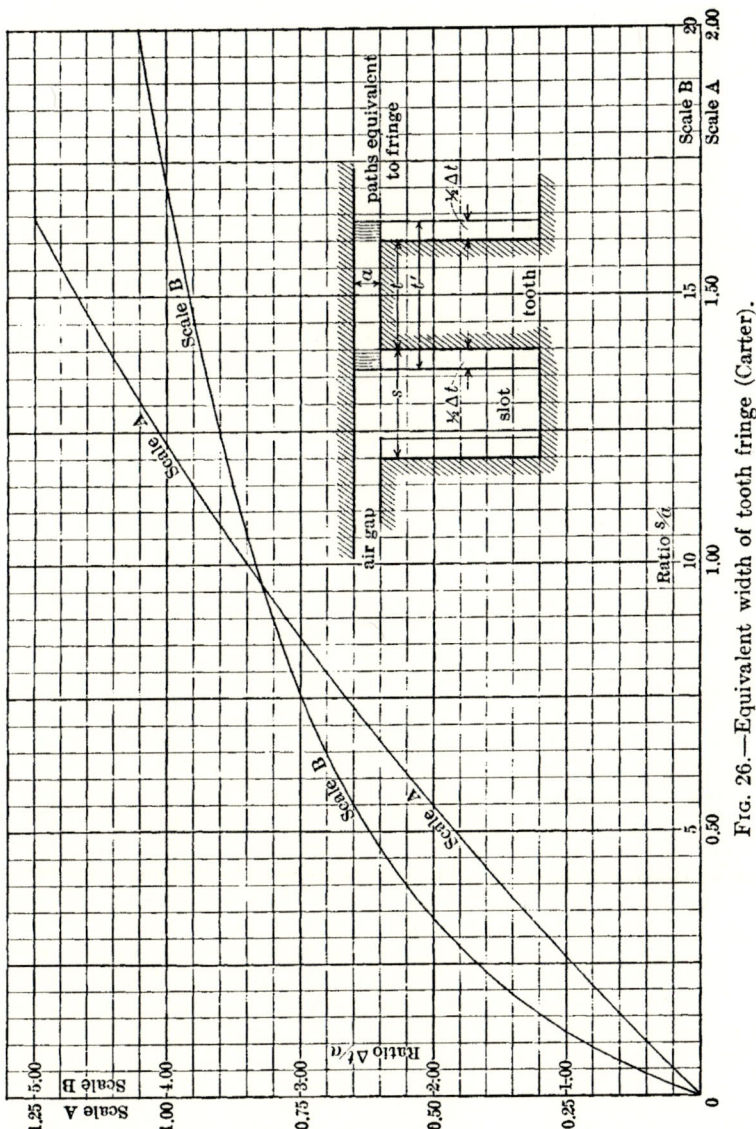

Fig. 26.—Equivalent width of tooth fringe (Carter).

Note: With semi-closed slots s and t are the dimensions at the armature periphery, and not those below the overhang.

flank surfaces of the two iron structures. This permeance is, however, very small, and has to be estimated empirically, if at all.

Strictly speaking, l_{eff} is different for each tooth, if the air-gap is variable, because the amount of fringing in the air-ducts and at the flanks is different. However, it is hardly worth the effort in ordinary cases to calculate l_{eff} for each tooth. It is sufficient to take an average l_{eff} for some intermediate value of the air-gap.

In some high-speed alternators, and usually in induction motors, air-ducts are provided in both the stationary and the revolving parts, in the same planes. The flux fringe in an air-duct is then of such a shape that the lines of force are parallel to one another in the middle of the air-gap, between the stator and the rotor. Therefore, when using the curve in Fig. 26 for such a case, the cylindrical surface midway between the stator and the rotor must be taken to correspond to that of the solid iron surface assumed in the deduction of the curve. Hence, $\frac{1}{2}a_x$ must be used instead of a_x in determining $\mathcal{H}$.

Having calculated the permeances of the several paths per pole pitch the total permeance of the air-gap is found as their sum, or

$$\mathcal{P}_a = \Sigma \mathcal{P}_{at} + 2\mathcal{P}_{fringe}. \quad \quad \quad \quad (49)$$

Then, the required number of ampere-turns is determined from eq. (44). The method gives quite correct results, especially with some experience in estimating the permeances of irregular paths. Each designer usually modifies slightly the empirical factors which are indispensable in this method, and devises short cuts good for the particular kind of machine in which he is interested.

Instead of calculating the permeance of each tooth separately, some engineers replace the actual variable air-gap by an equivalent constant air-gap a_{eq}, either by the judgment of the eye, or as in prob. 18 above. The actual peripheral length of the pole arc is increased by from one to one and one-half a_{eq} on each side to take into account the fringing at the pole-tips. This gives the number of teeth under the pole. The permeance of each tooth is calculated from eq. (48) for $a_x = a_{eq}$, and is then multiplied by the number of teeth. With some practice, one can obtain in this manner quite accurate results at a considerable saving in time.

The method outlined above is not directly applicable to induction machines which have slotted cores on both sides of the air-gap (Fig. 23): At each instant some stator teeth are opposite rotor teeth, others bridge over some rotor slots, and *vice versa*. The amount of overlap varies from instant to instant, causing periodic fluctuations in the air-gap reluctance.

Assume first that both the stator and the rotor have smooth surfaces facing the air-gap. Let the permeance of such a machine be $\mathcal{P}_s$. If now the armature be slotted, the cross-section of the paths in the air-gap (neglecting the fringe) is reduced in the ratio t_1/λ_1 where t_1 and λ_1 are the stator tooth width and tooth pitch respectively. The permeance $\mathcal{P}_s$ is also reduced in the same ratio. Let the rotor be also provided with slots; the *average* cross-section of the path is thereby further reduced in the ratio (t_2/λ_2), where t_2 and λ_2 are the tooth width and the tooth pitch on the surface of the rotor. Thus, disregarding the spread of the flux, the average air-gap permeance of an induction motor is

$$(\mathcal{P}_a) = (t_1/\lambda_1)(t_2/\lambda_2)\,\mathcal{P}_s,$$

the symbol $\mathcal{P}_a$ being put in parentheses to indicate that a further correction for the tooth fringe is necessary.[1]

In order to take the fringe into consideration, an empirical correction is made in this formula. Namely, it is assumed that the actual permeance of the fringes of the stator teeth is the same as if the rotor had a smooth core, and *vice versa*. Accordingly, in the preceding formula, the values t_1 and t_2 of the tooth widths are corrected for the fringe, using Carter's curve (Fig. 26). The formula becomes then

$$\mathcal{P}_a = (t_1'/\lambda_1)(t_2'/\lambda_2)\mathcal{P}_s, \quad \ldots \ldots \ldots \quad (50)$$

where t_1' and t_2' are the corrected widths of the stator and rotor teeth respectively. This formula has been found to be in a satisfactory agreement with experimental results.[2]

---

[1] For a more rigorous proof of this formula see C. A. Adams, "A Study in the Design of Induction Motors," *Trans. Amer. Inst. Electr. Engs.*, Vol. 24 (1905), p. 335.

[2] T. F. Wall, The Reluctance of the Air-gap in Dynamo-machines, *Journ. Inst. Electr. Engrs.* (British), Vol. 40 (1907–8), p. 568. E. Arnold in his *Wechselstromtechnik*, Vol. 5, Part 1, pp. 42, 43, calculates the value of k_a for an induction machine in a somewhat different way. With open slots in the stator, Arnold's method gives lower values of k_a than they are in reality. See Hoock and Hellmund, Beitrag zur Berechnung des Magnetizierungs-

Referring to Art. 36, eq. (50) may be interpreted as follows: Let k_a' be the air-gap factor for the slotted stator and a smooth-body rotor; let k_a'' be the same factor for the slotted rotor and a smooth-body stator. Then the air-gap factor of the actual machine

$$k_a = k_a' \times k_a''. \qquad \qquad (51)$$

In an induction motor the magnetic flux is distributed in the air-gap approximately according to the sine law, due to the distributed polyphase windings. Therefore the value of M determined from eq. (44) gives only the average value of the m.m.f. required for the air-gap. With a sine-wave distribution of the flux the maximum m.m.f. is $\pi/2$ times larger than the average value.

Prob. 20. What is the permeance of the air-gap of a 16-pole direct-current machine, the armature of which has a diameter of 250 cm. and is provided with 324 slots, 12 by 15 mm.? The gross length of the armature is 23 cm., and it is provided with three ventilating ducts, 10 mm. wide each. The axial length of the poles is 21.5 cm. The pole shoes cover 65 per cent of the periphery, and are not chamfered. The length of the air-gap is 10 mm.   Ans. About 900 perm.

Prob. 21. The machine mentioned in problem 14 has 120 slots, 3 by 6.5 cm. The pole-shoes are shaped according to the arc of a circle of a radius equal to 90 cm. and subtending 36 degrees on that circle; the pole-tips are formed by quadrants of a radius equal to 2.5 cm. Check the value of the field current (45 amp.) given in problem 15, by the method of equivalent permeances.

Prob. 22. What is the maximum m.m.f. across the air-gap of an induction motor, if the gross average flux density in the air-gap (total flux divided by the gross area of the air-gap, not including the vents) is 3 kl./sq.cm., and the clearance is 1.2 mm.? The bore is 64 cm.; the stator is provided with 48 open slots, 22 by 43 mm. The rotor has 91 half-closed slots, the slot opening being 3 mm. The machine has a vent 7 mm. wide for every 9 cm. of the laminations.

Ans. 820 amp. turns.

Prob. 23. Show that $\mathcal{J}l/a = 1.2 + 2.93 \log (s/2a)$, if the fringing lines of force are assumed to be concentric quadrants (Fig. 27, to the left) with the points c as the center; the average length of path in the part bcc' is estimated to be equal to $1.2a$, and the average width $0.72a$. Hint: The permeance of an infinitesimal tube of force of a radius x and of a width dx is $\mu \cdot dx/(\frac{1}{2}\pi x)$. Integrate this expression between the limits

of a and $\frac{1}{2}s$.   See Adams, *loc. cit.*, p. 332.   The formula can be used only
when s is larger than $2a$.

Prob. 24.   Show that $\mathit{J}t/a = 2.93 \log (1 + \frac{1}{4}\pi s/a)$, if the fringing lines
of force are assumed to consist of concentric quadrants (Fig. 27, to

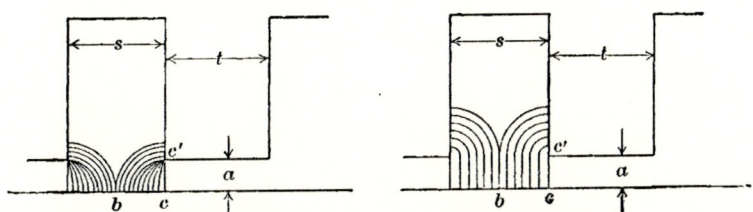

FIG. 27.—Two simplified paths for the fringing flux.

the right), with the point c' as a center, continued as straight lines.
See Arnold, *Die Gleichstrommachine*, Vol. 1, p. 269.

Prob. 25.   Show that formula (50) applies to synchronous and direct-
current machines with salient poles as well, if t_2' is the width of the
pole shoe, corrected for the fringe, and λ_2 is the pole pitch.

CHAPTER VI

EXCITING AMPERE-TURNS IN ELECTRICAL MACHINERY—(*Continued*)

38. The Ampere-turns Required for Saturated Teeth. The teeth and the slots of an armature, under the poles, are magnetically in parallel (Fig. 24); hence, part of the flux passes from the pole into the armature core through the slots between the teeth. But, with a moderate saturation in the teeth, say below 18 kilolines per square centimeter, the amount of the flux which passes through the slots is altogether negligible. If the taper of the teeth is slight, the required ampere-turns are found for the average flux density in the tooth, taking the value of H from the curves in Fig. 3.

Should the taper of the teeth be considerable, as is the case in revolving armatures of small diameter, the flux density should be determined in say three places along the tooth, viz., at the root, in the middle part, and at the crown. Let the corresponding values of magnetic intensity from the magnetization curve of the material be H_0, H_m, and H_1. Assuming H to vary along the tooth according to a parabolic law, we have, according to Simpson's rule in the first approximation, that the average intensity over the tooth is

$$H_{ave} = \tfrac{1}{6}H_0 + \tfrac{2}{3}H_m + \tfrac{1}{6}H_1. \quad \ldots \ldots (52)$$

If a greater accuracy is desired, the values of H can be determined for more than three cross-sections of the tooth and Simpson's rule applied.[1] For instance, let the length be divided into n equal

---

[1] A designer who has to calculate ampere-turns for teeth frequently will save time by plotting curves for the average H against the flux density B_0 at the root of the teeth. Each curve would be for one taper, and these curves would cover the usual range of taper in the teeth. See A. Miller Gray "Magnetomotive Force in Non-uniform Magnetic Paths," *Electrical World*, Vol. 57 (1911), p. 111.

parts, where n is an even number.   Then, we have that

$$H_{ave} = \frac{1}{3n}[(H_0+H_n)+4(H_1+H_3+\ldots+H_{n-1})$$
$$+2(H_2+H_4+\ldots+H_{n-2})].\quad . \quad (53)$$

When the flux density in the teeth is considerable, say between 18 and 24 kilomaxwells per square centimeter, an appreciable part of the total flux passes through the slots between the teeth, also through the air-ducts, and in the insulation between the laminations.   Dividing, therefore, the flux per tooth pitch by the net cross-section of the tooth, one gets only the so-called *apparent flux density* in the tooth, which density is higher than the *true density*. With highly saturated teeth, a small difference in the estimated flux density makes an appreciable difference in the required number of ampere-turns; it is therefore of importance to know how to determine the true density in a tooth, knowing the apparent density.

Consider first the case of a machine with a large diameter, in which the taper of the teeth can be neglected.   Assume the concentric cylindrical surfaces at the tips and at the roots of the teeth to be equipotential surfaces, and the lines of force to be all parallel to each other, in the slots as well as in the iron.   In reality, some lines of force enter the teeth on the sides of the slots (Fig. 24), so that the foregoing assumptions are not quite correct; but they are the simplest ones that can be made.   Any other assumptions would lead to calculations too complicated for practical use.

Let B_{real} be the true flux density in the iron of the tooth, and let B_{app} be the apparent flux density in the tooth under the assumption that no flux passes through the slots, air-ducts, or insulation between the laminations.   Then, denoting the actual flux density in the air by B_a, we have the following expression for the total flux per tooth pitch:

$$A_i B_{app} = A_i B_{real} + A_a B_a,$$

where A_i and A_a are the cross-sections in square centimeters of the paths per tooth pitch, in the iron and air respectively.   Since the iron and the air paths are of equal length, and are in parallel, the m.m.f. gradient is the same in both.   Let H be this gradient, in kiloampere-turns per centimeter. Then, if all the flux densities are in kilomaxwells per square centimeter,

$$B_a = 1.25H.$$

Substituting this value of B_a into the preceding equation we obtain, after division by A_i,

$$B_{app} = B_{real} + 1.25(A_a/A_i)H. \qquad \ldots \qquad (54)$$

The ratio A_a/A_i can be expressed through the dimensions of the machine as follows: $A_i = tl_n$ where t is the width of the tooth, and l_n is the net axial length of the laminations, without the air-ducts and insulation. $A_a = \lambda l_g - tl_n$, where λ is the tooth pitch (Fig. 20), and l_g is the gross length of the armature core. Hence,

$$A_a/A_i = (\lambda/t)(l_g/l_n) - 1. \qquad \ldots \qquad (55)$$

In eq. (54) the flux density B_{app} and the ratio A_a/A_i are known in any particular case, and the problem is to find B_{real} and H. The other equation which connects B_{real} and H is the magnetization curve of the material, and the problem can be solved in a similar manner to problem 11 in chapter II (see also problem 4 below).

Professional designers use curves like those shown in Fig. 28, which give directly the relation between B_{app} and B_{real} within the range of values of A_a/A_i which occur in practice. The curves are plotted point by point by assuming certain values of B_{real} and calculating the corresponding B_{app} from eq. (54). For instance, for $B_{real} = 24$, the saturation curve shown in Fig. 28 gives $H = 1.33$, so that for $A_a/A_i = 2$, we have: $B_{app} = 24 + 1.25 \times 2 \times 1.33 = 27.33$. This determines one point on the curve marked "Ratio of air to iron = 2." In using these curves one begins with the known value of B_{app} on the lower axis of abscissæ, and follows the ordinate to the intersection with the curve for the desired ratio A_a/A_i; this gives the value of B_{real}. By following the horizontal line from the point so located to the intersection with the B-H curve, the corresponding value of H is read off on the upper axis of abscissæ.

The curves in Fig. 28 are completely determined by the shape of the B-H curve, so that, if the material to be used for the armature core differs considerably from that assumed in Fig. 28, new curves of B_{real} versus B_{app} ought to be plotted, or else the method

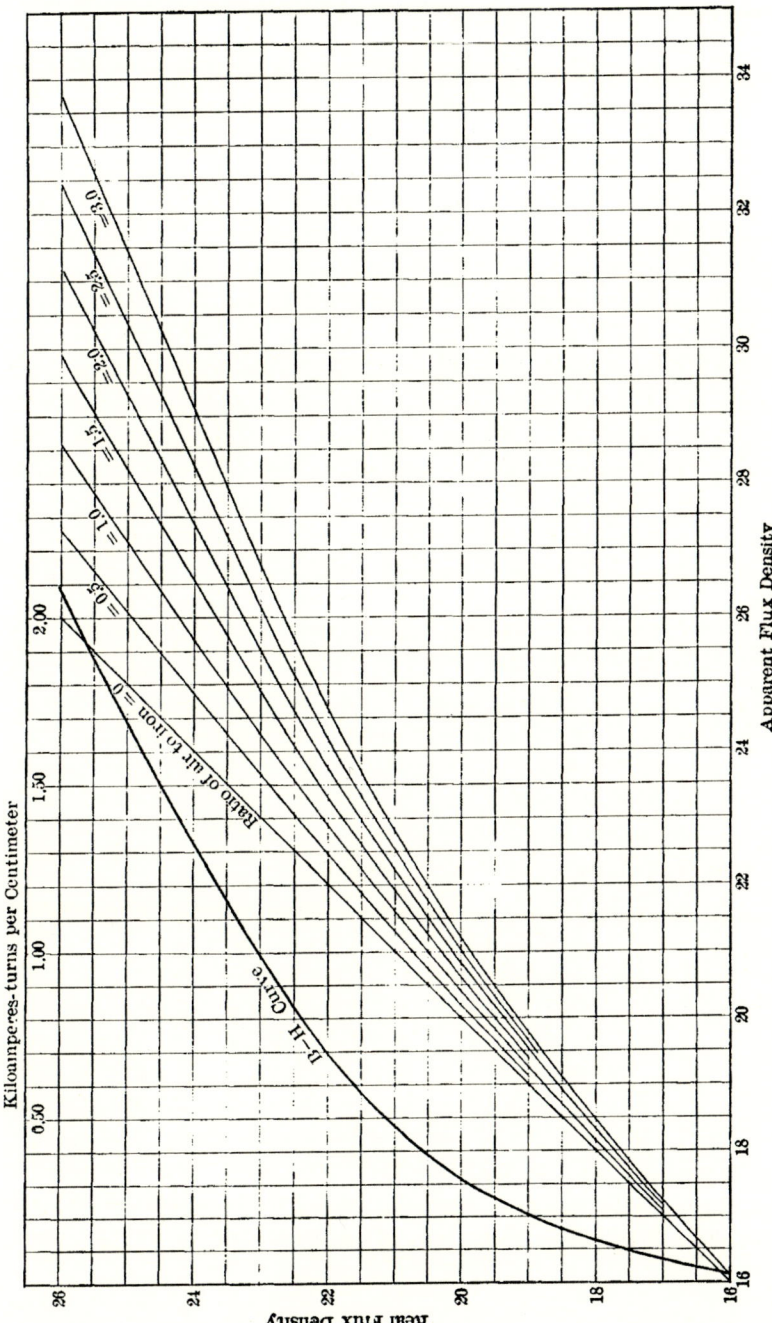

Fig. 28.—The Relation between the real and the apparent flux density in armature teeth.

may be used which is suggested in problem 4 below.   A comparison of the B-H curve with those in Fig. 3 shows that a much better quality of steel is presupposed in Fig. 28.   Such is usually the case when it is desired to employ highly saturated teeth, for otherwise it might be practically impossible to get the required flux.   The curves in Fig. 3 refer to an average quality of electrical steel.

Formula (54) and the curves in Fig. 28 presuppose that the teeth have no taper, or that the taper is negligible.   If the taper of the teeth is quite considerable the tooth and the slot are divided by equipotential cylindrical surfaces into two or more parts, and H is determined for each part.   The effective value of H is calculated according to Simpson's rule, using formula (52) or (53).

Prob. 1.   A four-pole direct-current armature has the following dimensions: diameter 45 cm.; gross length of core 20 cm.; two air-ducts 7 mm. each; 67 open slots 1 by 3 cm.   The poles are of such a shape that the flux per pole is carried uniformly by 11.5 teeth.   How many ampere-turns per pole are required for the teeth when the flux per pole is 3 megalines?   Use the saturation curve for carbon-steel laminations in Fig. 3, and neglect the effect of slight saturation at the roots of the teeth.   Ans.   148 amp.-turns.

Prob. 2.   How many ampere-turns are required in the preceding problem when the flux per pole is 4.4 megalines?
          Ans.   Between 2750 and 2800.

Prob. 3.   The machine, in problem 22 of the preceding chapter, had a gross average flux density in the air-gap of 3 kl./sq. cm.   The bore was 64 cm.   The stator was provided with 48 slots 22 by 43 mm.   The machine has a vent 7 mm. wide for every 9 cm. of the laminations. What is the maximum m.m.f. required for the stator and rotor teeth, if the size of each of the 91 rotor slots is 14 by 30 mm. below the overhang? Take an overhang of 2.8 mm.   Ans.   About 220 amp.-turns.

Prob. 4.   Instead of drawing the curves shown in Fig. 28, the relation between B_{real} and B_{app} can be found by the following construction: Disregard the lower scale marked " Apparent Flux Density "; extend the left-hand scale to the division 34 and mark the scale " Real and Apparent flux density."   Cut out a strip of paper, and copy the left-hand scale on the left-hand edge of the strip.   On the right-hand edge of the strip mark the scale for A_a/A_i as follows: division 26 of the flux density to correspond with zero, division 27 with 0.4, division 28 with 0.8, etc. Apply the left-hand edge of the strip to division 2.0 on the upper horizontal scale, and to division 26 on the lower horizontal scale.   Move the strip up and down until the upper horizontal scale coincides with the desired value of A_a/A_i marked on the strip.   Lay a straightedge on the divisions of the two vertical scales corresponding to the given apparent flux density.   The intersection of the straightedge with the B-H curve will give the required values of B_{real} and H. Check this construction for a few points with the values obtained from the curves, and give a general proof. Hint: This construction amounts to considering the B-H curve and Eq. (54) as two simultaneous equations with two unknown quantities B_{real} and H. See problem 11 in chapter II, Art. 13.

39. The Ampere-turns for the Armature Core and for the Field Frame. In many machines the m.m.f. required for the air-gap and the teeth are large as compared to those required for the armature core and the field frame; in such cases the latter are either altogether neglected, or are estimated roughly, by increasing the ampere-turns calculated for the rest of the magnetic circuit by say five or ten per cent. Where this is not permissible, the usual procedure is to estimate the maximum flux density in the core or frame under consideration and to measure from the drawing of the machine the length of the average path of the lines of force in it. The assumption is made that the same flux density is maintained on the whole length of the path, and the required ampere-turns are calculated from the magnetization curve of the material (Figs. 2 and 3). While the ampere-turns determined in this way are usually larger than those actually required, the method is permissible if the total amount of the m.m.f. for the parts under consideration is small as compared to the total m.m.f. of the magnetic circuit. If a greater accuracy is desired, the path is subdivided into two or more parts in series, and the average density determined for each part; and then the ampere-turns required for each part are added.

The tendency now is to increase the flux density in the armature cores of alternators and induction motors so as to reduce the size of the machine. This is made possible through a better quality of laminations, which show a smaller core loss, and also through the use of a more intensive ventilation. With these high densities and with the comparative large values of the pole pitch necessary in high-speed machinery, the ampere-turns for the core constitute an appreciable amount of the total m.m.f. of the machine, and it is therefore desirable to calculate them more accurately.

The flux density in the core is a minimum opposite the center of a pole, and is a maximum in the radial plane midway between two poles (Fig. 15). At each point the flux density has a tangential and a radial component. The latter is comparatively small and can be neglected; the tangential component can be assumed to vary according to the sine law, being zero opposite the center of the pole and reaching its maximum between the poles. With these assumptions, knowing the maximum flux density in the core, the flux density at all other points is calculated, and the corresponding values of H are determined from the B-H curve of the material. The average value of H for one-half pole pitch is then found by Simpson's rule, eqs. (52) and (53). With the sine-wave assumption, the average H depends only upon the maximum flux density, so that for a given material a curve can be compiled from the B-H curve, giving directly H_{ave} for different values of B_{max}.[1]

Should a still greater accuracy be required, the following method can be used: Draw the assumed or the calculated curve of the distribution of flux density in the air-gap, and indicate to your best judgment the tubes of force in the armature core, say for each tooth pitch. The flux in the radial plane midway between the two poles can be assumed to be distributed uniformly over the cross-section, and this fact facilitates greatly the determination of the shape of the tubes of flux. The m.m.f. required for each tube is calculated by dividing it into smaller tubes in series and in parallel; thus, either the average m.m.f. for the whole flux can be found, or the maximum m.m.f. for one particular tube.[2]

The frame to which the poles are fastened in direct-current and in synchronous machines is usually made of cast iron; in some cases the frame is made of cast steel; in high-speed synchronous machines the revolving field is made of forged steel. The magnetomotive force required for such a frame is found in the usual way from the magnetization curve of the material, knowing the area and the average length of the path between two poles; the length is estimated from the drawing of the machine. In figuring out the flux density in a field frame one must not forget that (1) only one-half of the flux per pole passes through a given cross-section of the frame (Fig. 20); (2) the total flux in the frame and in the poles is larger than that in the armature by the amount of the leakage flux between the poles. This leakage is usually estimated in per cent

---

[1] This method is due to E. Arnold. See his *Wechselstromtechnik*, Vol. 5, (1909), part 1, p. 48.

[2] For details of this method see Hoock and Hellmund, Beitrag zur Berechnung des Magnetizierungsstromes in Induktionsmotoren. *Elektrotechnik und Maschinenbau*, Vol. 28 (1910), p. 743.

of the useful flux, from one's experience with previously built machines, or it can be calculated by the methods explained in the next article. Thus, a *leakage factor* of 1.20 means that the flux in the field poles is 20 per cent higher than that in the armature, the leakage flux constituting 20 per cent of the useful flux. The usual values of the leakage factor vary between 1.10 and 1.25, depending upon the proximity of the adjacent poles, the degree of saturation of the circuit, and the proportions of the machine.

The ampere-turns required for the pole-pieces are calculated in a similar way, assuming the whole leakage to take place between the pole-tips, so that the flux density in the pole-waist corresponds to the total flux, including the leakage flux. In exceptional cases of highly saturated pole-cores this method may be inadmissible, on account of too large a margin which it would give as compared to the ampere-turns actually required. In such cases part of the leakage may be assumed to be concentrated between some two corresponding points on the waists of two adjacent poles, or it may be assumed to be actually distributed between the two pole-waists. See probs. 9 and 10 in chapter II.

In some machines the joint between the pole and the frame offers a perceptible reluctance, like the joints in the transformer cores discussed in Art. 33. Some designers allow a certain fraction of a millimeter of air-gap to account for this reluctance, and add the number of ampere-turns required to maintain the flux in this air-gap to those for the pole-piece. The length of this equivalent air-gap is found by checking back no-load saturation curves obtained from experiment. As a usual rule, .t is advisable to increase the total calculated ampere-turns of the magnetic circuit by about 5 to 10 per cent. This increase covers uch minor points as the reluctance of the joints, omitted in calculations, as well as certain inaccurate assumptions; it also covers a possible discrepancy between the assumed and the actual permeability of the iron. With a liberally proportioned field winding and a proper regulating rheostat a designer can rest assured that the required voltage will be obtained, though possibly at a somewhat different value of the field current than the estimated one.

Prob. 5. The stator core of a six-pole induction motor has the following dimensions : bore 112 cm.; outside diameter 145 cm.; gross length 55 cm.; the slots are 2 cm. by 4.5 cm.; t' e machine is provided with

8 ventilating ducts 9 mm. wide each. What is the maximum m.m.f. required for the stator core per pole if the flux per pole is 0.15 weber?
Ans. 187 using Arnold's method.

Prob. 6. Draw a curve between the average H and maximum B in the core, assuming a sinusoidal distribution of the flux density in the tangential direction, for the carbon steel laminations in Fig. 3.
Ans. $H_{ave} = 26.8$ for $B_{max} = 18$.

Prob. 7. The cross-section of the cast-iron field yoke of a direct-current machine is 650 sq.cm.; the mean length of path in it between two consecutive poles is 85 cm. The length of the lines of force in each pole-waist is 21 cm.; its cross-section 420 sq.cm. The poles are made of steel laminations 4 mm. thick, so that the space lost between the laminations is negligible. The reluctance of the joint between a bolted pole and the yoke is estimated to be equivalent to 0.1 mm. of air. What is the required number of ampere-turns for the pole-piece and the yoke, per pole, when the useful flux of the machine is 5 megalines per pole? The leakage factor is estimated to be equal to 1.20.
Ans. About 930.

40. Magnetic Leakage between Field Poles.

It is of importance in modern highly saturated machines to know accurately the leakage flux between the poles, in order to estimate correctly the ampere-turns required for the field poles and the frame of the machine. Moreover, the design of the poles can be improved, knowing exactly where the principal leakage occurs and how it depends upon the proportions of the machine. The value of the leakage factor also affects the voltage regulation of the machine, because at full load the m.m.f. between the pole-tips has to be larger than at no-load, on account of the armature reaction.

For new machines of usual proportions the value of the leakage factor can be estimated from tests made upon similar machines. But in new machines of unusual proportions the designer has to rely upon his judgment, assisted if necessary by crude comparative computations of the permeance between adjacent poles. In this and in the next article some examples of such computations are given, not so much in order to give a definite method to be followed in all cases, as to show the student a possible procedure and to train his judgment in estimating the permeance of an irregular path.

Four principal paths of leakage can be distinguished between two adjacent poles (Fig. 29): (a) between the sides of the pole shoes which face each other; (b) between the sides of the pole cores (waists) parallel to the shaft of the machine; (c) between the flanks or sides of the pole shoes perpendicular to the shaft; (d) between the flanks of the pole cores. In the calculations which follow, the permeances are computed between a pole and the planes of symmetry, MN, between the two poles, the permeance of the other half of each path being the same. All these leakage paths are in parallel with respect to the pole, so that the total leakage

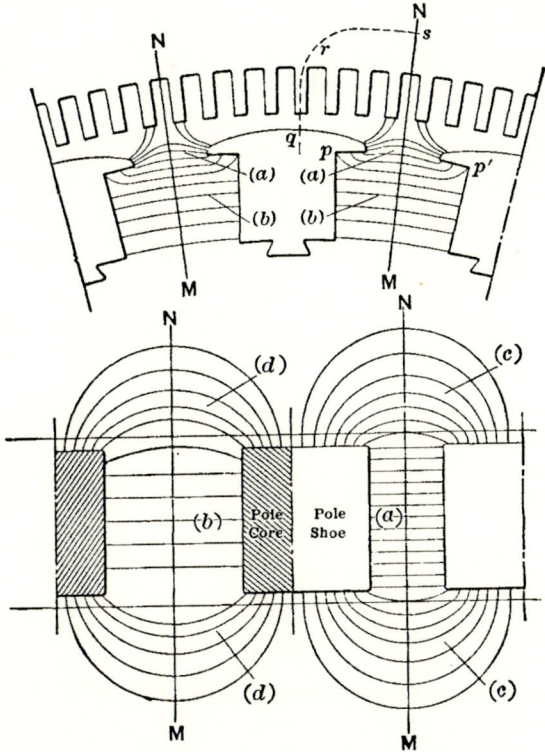

FIG. 29.—The leakage flux between field poles.

permeance is equal to their sum. Knowing this total permeance and the m.m.f. between the pole and the plane MN the leakage flux is found, and knowing this flux and the useful flux per pole the leakage factor is easily calculated.

We shall now estimate the permeances of each of the four above mentioned paths of leakage.

(a) *Between the adjacent pole-tips.* Estimate the average cross-section A of the path, in square centimeters, and the average length l between the pole-tip and the plane MN, being guided by Fig. 29. (Do not encroach upon the fringe to the armature.) Then the permeance of the path is, in perms,

$$\mathcal{P} = \mu A / l. \qquad \dots \dots \dots \quad (56)$$

If a greater accuracy is desired, subdivide the total path into smaller paths in series and in parallel, and calculate the permeance or the reluctance of each separately. Then the total permeance is found according to the well-known law of combination of reluctances and permeances in series and in parallel (Art. 9). When mapping out the lines of force in the air, begin them nearly at right angles to the surface of the pole (see Art. 41 a below) and draw them so as to make the total permeance of the path a maximum, that is, reducing as far as possible the length and increasing the cross-section of each elementary tube of flux. The medium may be said to be in a state of tension along the lines of force, and of compression at right angles to their direction, by virtue of the energy stored in the field. Hence, there is a tendency for the tubes of force to contract along their length and expand across their width.

(b) *Between the opposite pole-cores.* In this part of the leakage field each elementary concentric path is subjected to a different m.m.f., that between the roots of the poles being practically zero, while the m.m.f. between the points p and p' is equal to that between the pole-tips. In most cases it is permissible to consider the whole leakage flux as if passing through the whole length of the pole core, and then crossing to the adjacent poles at the pole-tips. Therefore, it is convenient to add the permeance between the pole cores to that between the pole-tips. But the average m.m.f. between the waists is only about one-half of that between the tips, so that the equivalent permeance between the pole cores, reduced to the total m.m.f., is equal to one-half of the actual permeance. If the actual permeance, calculated according to formula (56) is $\mathcal{P}$, the effective permeance is $\frac{1}{2}\mathcal{P}$. The average length and cross-section of the path are easily estimated from the drawing of the machine.

(c) *Between the flanks of the pole shoes.* The path extends indefinitely outside the machine, and the lines of force are twisted curves, so that it is difficult to estimate the permeance graphically. As a rough estimate, this permeance can be reduced to that of the path in the air between two rectangular poles of an electromagnet (Fig. 30). Assume the paths of the flux to consist of concentric quadrants with the centers at c and c', joined by parallel straight lines, and let the width of the poles in the direction perpendicular to the plane of the paper be h. Then the permeance of an infinitesimal layer of thickness dx, between one of the poles and the plane MN of symmetry, is

$$d\mathcal{P} = \mu dx.h/(\tfrac{1}{2}\pi x + l).$$

Integrating this expression between the limits o and b we find

$$\mathcal{P} = 1.84h \log (1.57b/l + 1) \text{ perms} \quad . \quad . \quad . \quad (57)$$

(compare with prob. 24 in Chapter V, Art. 37).

In applying this formula and Fig. 30 to the case of the flank leakage between the pole shoes, h is the average radial height of the pole shoe, b is equal to one-half the width of the pole shoe, and $2l$ is the distance between the two opposing pole-tips. While the method evidently gives only a crude approximation to the actual permeance, formula (57) at least fixes a lower limit to the permeance in question.

(d) *Between the flanks of the pole cores.* The conditions are similar to those under (c), so that the permeance is estimated again on the basis of formula (57). The sides of the two rectangles in Fig. 29 are not parallel to each other as in Fig. 30, but this difference is taken into account by mentally turning them into a parallel

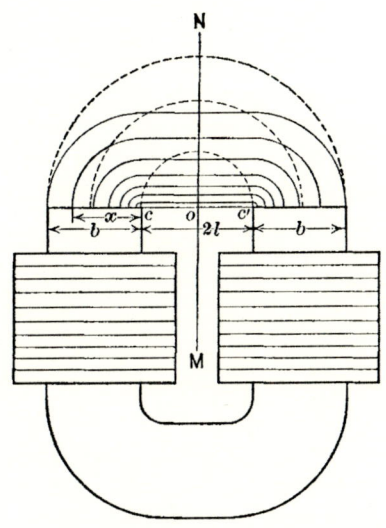

FIG. 30.—The magnetic path between the poles of an electromagnet.

position, and estimating the equivalent distance $2l$ between the edges of the opposing poles. The dimension h is in this case the radial height of the pole-waist, and b is one-half of the width of the pole-waist. The flank leakage is smaller than that between ·

the opposite side, so that one may be satisfied with a lesser degree of accuracy. The equivalent permeance, reduced to that between the pole-tips, is again equal to one-half the actual permeance, for the same reason as under (*b*) above.

The total leakage permeance between a pole and the two planes of symmetry is equal to the sum of the permeances calculated as above. In summing them up it will be seen from Fig. 29 that the permeances (*a*) and (*b*) must be taken twice, and also that (*c*) and (*d*) must be taken four times. The leakage flux per pole is obtained by multiplying the total leakage permeance by the m.m.f. between the pole-tip and the plane of symmetry. This m.m.f. is equal to that required to establish the useful flux, along the path *qrs*, through the air-gap and the armature of the machine, and consequently it is known before the pole-piece and the field winding are computed in detail. Knowing the leakage flux and the useful flux, the leakage factor is figured out according to the definition given above.

When calculating permeances as indicated above, one is advised to make liberal estimates of the same, for two reasons: In the first place, the true permeance of a path is always the largest possible, so that, whatever assumptions one makes, the calculated permeance comes out smaller than the actual. In the second place, in designing a new machine it is better to be on the safe side and rather underestimate than overestimate the excellence of the performance. Some writers give more elaborate rules and formulæ for the calculation of the leakage permeance which are useful in the design of machines of special importance.[1]

The leakage factor remains practically constant as long as the flux density in the armature core and teeth is moderate, so that the reluctance of the useful path *qrs* is nearly constant. This is because the reluctance of the leakage paths is constant, and, if the reluctance of the useful path is also constant, the useful flux and the leakage flux increase in the same proportion when the m.m.f. between the pole-tips is increased. When the armature iron is approaching saturation, the leakage factor increases with the field

[1] For a more detailed treatment of the leakage between poles see the following works: E. Arnold, *Die Gleichstrommaschine*, Vol. 1 (1906), pp. 284–294; Hawkins and Wallis, *The Dynamo*, Vol. 1 (1909), pp. 469–484; Pichelmayer, *Dynamobau* (1908), pp. 127–131; Cramp, *Continuous-Current Machine Design* (1910), pp. 42–47 and 226–230.

current, because the leakage flux increases the more rapidly than the useful flux. This increase is partly offset by the fact that the pole-tips also become gradually saturated by the leakage flux, so that the leakage factor does not increase as rapidly as it would otherwise. The practical point to be observed is, that for the higher flux densities, if accuracy is desired, the leakage should be estimated separately for a few points on the no-load saturation curve.

For a given terminal voltage, the leakage factor of a machine is somewhat higher at full-load than at no-load, because the required m.m.f. between the pole-faces is higher, due to the armature reaction and to the voltage drop in the armature. In comparatively rare cases, when the armature reaction assists the field m.m.f., for instance, in the case of an alternator supplying a leading current, the leakage factor decreases with the increasing load. The following example illustrates the influence of the load upon the value of the leakage factor.

Let the useful flux per pole in an alternator, at the rated voltage and at no-load, be 5 megalines, and let 6000 amp.-turns per pole be required for the air-gap and the armature core. Let the permeance of the leakage paths between a pole and the neutral planes be 120 perms, so that the leakage flux is 0.72 megaline, and the leakage factor is $(5.00 + 0.72)/5.00 = 1.14$. Let a useful flux of 5.5 megalines be required at the same voltage and at full load, an increase of 10 per cent being necessary to compensate for the internal drop of voltage due to the armature impedance. If the teeth and the armature core were not saturated at all, an m.m.f. of 6600 amp.-turns would be required. In reality, the m.m.f. is higher, say 7500 amp.-turns. Let the armature reaction be equal to 1500 demagnetizing ampere-turns per pole. To compensate for its action, 1500 additional ampere-turns are required on each field coil. Thus, the difference of magnetic potential between a pole-tip and the adjacent plane of symmetry MN (Fig. 29) is now 9000 amp.-turns, and the leakage flux is increased to 1.08 megalines. Therefore, the leakage factor at full load is $(5.50 + 1.08)/5.50 = 1.20$. Similar relations hold for the direct-current machines.

In calculating the performance of a synchronous or a direct-current machine one has to use the relation between the field current and the voltage induced in the armature. Ordinarily, the no-load saturation curve is used for this purpose, assuming that the leakage factor is the same at full load as at no load. However careful designers sometimes plot a separate curve, using a higher leakage factor, for use at full load.

Prob. 8. Assume in the illustrative example given in the text the armature current to be leading, so that the voltage drop in the armature is negative and the armature reaction strengthens the field. Show that with the same value of the armature current the leakage factor is between 1.09 and 1.10.

Prob. 9. Draw rough sketches of the magnetic circuits of two machines, one possessing such proportions, number and shape of poles as to give a particularly low leakage factor, the other markedly deficient in this respect.

Prob. 10. Calculate the leakage factor and the leakage permeance per pole of a six-pole turbo-alternator of the following dimensions: The bore is 1.2 m.; the axial length of the poles 0.6 m.; minimum air-gap 1 cm.; radius of pole-shoe arc 45 cm.; total height of the pole 26 cm.; the height of the pole-waist 21 cm.; the breadth across the pole-waist 25 cm.; that across the pole-tips 36 cm. The reluctance of the useful path in the air-gap and in the armature is estimated to be about 0.57 millirel per pole, at normal no-load flux.   Ans. 1.114; about 200 perms.

Prob. 11. The leakage factor of the machine specified in the preceding problem was found from an experiment to be 1.13, at no-load, when the total flux per pole was 20.35 megalines. What is the true leakage permeance if 10 kiloampere-turns per pole were required at that flux for the air-gap and the armature?   Ans. 234 perms.

Prob. 12. The machine specified in the two foregoing problems requires at full load 20 per cent more ampere-turns for the air-gap and armature, on account of the induced voltage being 12 per cent higher than at no-load. The armature reaction amounts to 4000 demagnetizing ampere-turns per pole. What is the leakage factor at full load, according to the calculated leakage permeance and according to that obtained from the test?   Ans. 1.16; 1.19.

Prob. 13. A closed electric circuit consisting of a battery and of a bare conductor is immersed in a slightly conducting liquid, so that part of the current flows through the liquid. Indicate the common points and the difference between this arrangement and a magnetic circuit with leakage. Using the electrical analogy, show that armature reaction increases the leakage factor; also explain the fact that, in order to compensate for the action of M demagnetizing ampere-turns on the armature, more than M additional ampere-turns are required on the pole-pieces.

Prob. 14. In some books the permeance between two pole-faces (Fig. 30) is calculated by assuming the lines of force to be concentric semicircles as shown by the dotted lines. Show that such a permeance is smaller than that according to formula (57) and therefore should not be used. Hint: Compare the lengths of two corresponding lines of force.

Prob. 15. Let AB and CD (Fig. 31) represent the cross-sections of two opposite pole-faces of an electromagnet, inclined at an angle 2θ to one another. Show that of the three assumptions with regard to the shape of the lines of force in the air between the poles (a) is more correct than (b) and (b) is more correct than (c); in other words, the assumption (a) gives a higher permeance than (b) or (c). Hint: $\tan \theta > \theta > \sin \theta$.

Prob. 16. Show that the permeance according to Fig. $31a$, between one of the faces and the plane MN of symmetry, is equal to

$$(\mu h/\theta)Ln(\theta w/l + 1),$$

where h is the width of the pole-faces perpendicular to the plane of the paper, and that formulæ (56) and (57) are special cases of it.

Prob. 17. The formula given in the preceding problem is deduced under the assumption that the same m.m.f. is acting on all the lines of

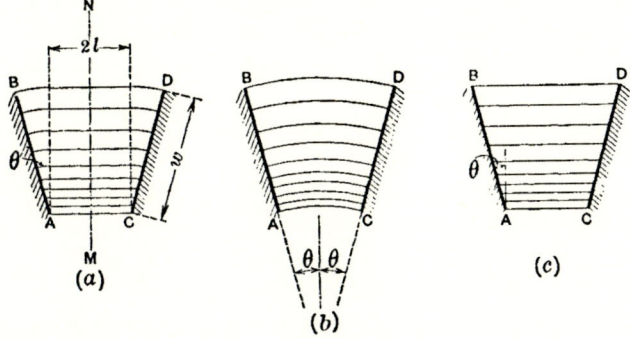

Fig. 31.—The magnetic paths between the poles of an electromagnet (three assumptions).

force. Let now Fig. $31a$ represent a cross-section of two opposing pole cores in an electric machine, the m.m.f. between A and C being zero, and uniformly increasing to a value M between the points B and D. Show that the equivalent permeance of the path, referred to the m.m.f. M is equal to $(\mu h/\theta)[1-(l/\theta w)Ln(\theta w/l + 1)]$.

Note. If it is desired to use regularly the foregoing formula in estimating the leakage factor, the values of the expression in the brackets [ ] can be plotted as a curve for the values of $(l/\theta w)$ as abscissæ. Similar formulæ can be deduced and curves plotted for the permeance of the flank leakage between adjacent poles. The paths of the lines of force over the poles can be assumed to be concentric quadrants and between the poles to have a shape similar to that indicated in Fig. $31a$.

41. The Permeance and Reluctance of Irregular Paths. In using the methods described above for the calculation of the ampere-turns for the air-gap, the teeth, and the cores, and in estimating the leakage factor, the reader has seen the difficulties involved in the computation of the permeance of an irregular path.

In the parts of a magnetic field not occupied by the exciting windings, the general principle applies that the lines of force and the equipotential surfaces assume such shapes and directions that the total permeance becomes a maximum, or the reluctance a minimum. When this condition is fulfilled, the energy of the magnetic field becomes a maximum, as is explained in Art. 57.

When the field needs to be considered in two dimensions only, that is, in the case where we have long cylindrical surfaces the properties of conjugate functions can be used for determining the equations of the lines of force and of the equipotential surfaces; see the references in Art. 37 above. However, the purely mathe- matical difficulties of the method are such as to make the analytical calculation of permeances feasible in the simplest cases only.

In most practical cases, especially in three-dimensional prob- lems, recourse must be had to the graphical method of trial and approximation, in order to obtain the maximum permeance. The field is mapped out into small cells by means of lines of force and equipotential surfaces, drawing them to the best of one's judgment; the total permeance is calculated by properly com- bining the permeances of the cells in series and in parallel. Then the assumed directions are somewhat modified, and the permeance is calculated again, etc., until by successive trials the positions of the lines of force are found with which the permeance becomes a maximum.

The work of trials is made more systematic by following a pro- cedure suggested by Lord Rayleigh. Imagine infinitely thin sheets of a material of infinite permeability to be interposed at intervals into the field under consideration, in positions approximately coinciding with the equipotential surfaces. If these sheets exactly coincided with some actual equipotential surfaces, the total permeance of the paths would not be changed, there being no tendency for the flux to pass along the equipotential surfaces. In any other position of the infinitely conducting sheets, the total permeance of the field is increased, because through these sheets the flux densities become more uniformly distributed. Moreover, these sheets become new equipotential surfaces of the system, because no m.m.f. is required to establish a flux along a path of infinite permeance. Thus, by drawing in the given field a system of surfaces approximately in the directions of the true equipotential surfaces, and assuming these arbitrary surfaces to be the true equipotential surfaces, the true reluctance of the path is reduced. In other words, by calculating the reluctances of the laminæ between the " incorrect " equipotential surfaces and adding these reluctances in series, one obtains a reluctance which is lower than the true reluctance of the path. This gives a lower limit for the required reluctance (or an upper limit for the permeance) of the path.

Imagine now the various tubes of force of the original field wrapped up in infinitely thin sheets of a material of zero permeability. This does not change the reluctance of the paths, because there are no paths between the tubes. But if these wrappings are not exactly in the direction of the lines of force, the reluctance of the field is increased, because the densities become less uniform, the non-permeable wrappings forcing the lines of force from their natural positions. Thus, by drawing in a given field a system of surfaces approximately in the directions of the lines of force, calculating the reluctances of the individual tubes, and adding them in parallel, a reluctance is obtained which is higher than the true reluctance of the path. This gives an upper limit for the reluctance (or a lower limit for the permeance) of the path under consideration.

Therefore, the practical procedure is as follows: Divide the field to the best of your judgment into cells, by equipotential surfaces and by tubes of force, and calculate the reluctance of the field in two ways: first, by adding the cells in parallel and the resultant laminæ in series; secondly, by adding the cells in series and the resultant tubes in parallel. The first result is lower than the second. Readjust the position of the lines of force and of the equipotential surfaces until the two results are sufficiently close to one another; an average of the two last results gives the true reluctance of the field.

One difficulty in actually following out the foregoing method is that the changes in the assumed directions of the field that will give the best result are not always obvious. Dr. Th. Lehmann has introduced an improvement which greatly facilitates the laying out of a field.[1] We shall explain this method in application to a two-

[1] " Graphische Methode zur Bestimmung des Kraftlinienverlaufes in der Luft "; *Elektrotechnische Zeitschrift*, Vol. 30 (1909), p. 995.

dimensional field, though theoretically it is applicable to three-dimensional problems also. According to Lehmann, lines of force and level surfaces are drawn at such distances that they enclose cells of equal reluctance. Consider a slice, or a cell, in a two-dimensional field, ν centimeters thick in the third dimension (where $\nu = 1/\mu$), and of such a form that the average length l of the cell in the direction of the lines of force is equal to its average width w in the perpendicular direction. The reluctance of such a cell is always equal to one rel, no matter whether the cell itself is large or small. This follows from the fundamental formula for the reluctance, which in this case becomes $\mathcal{R} = \nu l/(\nu \times w) = 1$.

The judgment of the eye helps to arrange cells of a width equal to the length, in the proper position with respect to each other and to the adjoining iron; the next approximation is apparent from the diagram, by inspecting the lack of equality in the average width and length of the cells. Lord Rayleigh's condition is secured by this means, since the combination of cells of equal reluctance leads to but one result, whether they are combined first in parallel or first in series. After a few trials the space is properly ruled, and it simply remains to count the number of cells in series and in parallel. Dr. Lehmann shows a few applications of his method to practical cases of electrical machinery, and the reader is referred to the original article for further details.

The foregoing methods apply only to the regions outside the exciting current, because only in such parts of the field the maximum permeance corresponds to the maximum stored electromagnetic energy. Within the space occupied by the exciting windings the condition for the maximum of energy is different (see Art. 57), and is of a form which hardly permits of the convenient application of a graphical method. However, in most practical cases the directions of the lines of force within the exciting windings are approximately known a priori; or else, the windings themselves can be assumed, for the purposes of computation, to be concentrated within a very small space. For instance, the field winding can be assumed to consist of an infinitely thin layer close to the pole-waist. Then the condition that the permeance is a maximum is fulfilled in practically the whole field, and the field is mapped out on this basis.

Prob. 18. Sketch the field between the armature and a pole-piece or some proportion of tooth, slot, and air-gap and determine the lower and upper limits of the reluctance by Lord Rayleigh's method.

Prob. 19. For some ratio of slot width to air-gap draw the tooth-fringe field to the perpendicular surface of the pole, adjust the number and spacing of the lines of force by Dr. Lehmann's method, and see how closely you can check the corresponding point on Carter's curve (Fig. 26).

Prob. 20. From the given drawing of a machine, determine the permeance of the fringe from the pole-tip to the armature by Lehmann's method; consult, if necessary, Dr. Lehmann's original article.

Prob. 21. Map out the leakage field between the opposing pole-tips and cores of a given machine, and determine its equivalent permeance by Lehmann's method, assuming the field coils to be thin and close to the core.

41a. The Law of Flux Refraction. When mapping out a field in air, the lines of force must be drawn so as to enter the adjoining iron almost normally to its surface, even if they are continued in the iron almost parallel to its surface. The lines of force change their direction at the dividing surface suddenly (Fig. 32), and in so doing they obey the so-called *law of flux refraction;* namely,

$$\tan \theta_i / \tan \theta_a = \mu_i / \mu_a \quad . \quad . \quad . \quad . \quad . \quad (58)$$

Since μ_a is many times smaller than μ_i, the angle θ_a is usually very small, unless θ_i is very nearly 90 degrees. It may be said in general that the lower the permeability of a medium the nearer the lines of force are to the normal at its limiting surfaces. In this way, the path between two given points is shortened in the medium of lower permeability and is lengthened in the medium of higher permeability. Thus, the total permeance of the circuit is made a maximum.

To deduce the above-stated law of refraction, consider a tube of flux between the equipotential surfaces *ab* and *cd*, the width of the path in the direction perpendicular to the plane of the paper being one centimeter. Let B_i and B_a be the flux densities, and H_i and H_a the corresponding magnetic intensities in the two media. Two conditions must be satisfied, namely, first, the drop of m.m.f.

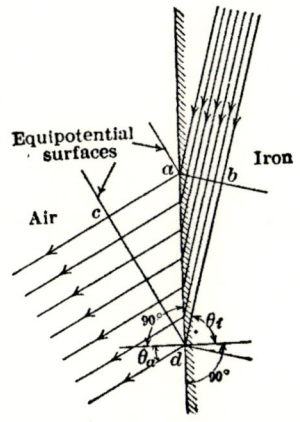

FIG. 32.—The refraction of a flux.

along ac is the same as that along bd, and secondly, the total flux through cd is equal to that through ab. Or

$$H_a \cdot \overline{ac} = H_i \cdot \overline{bd},$$

and

$$B_a \cdot \overline{cd} = B_i \cdot \overline{ab}.$$

Dividing one equation by the other, and rearranging the terms, eq. (58) is obtained.

Prob. 22. Show that the total refraction which is in some cases experienced by rays of light is impossible in the case of magnetic lines of force.

Prob. 23. Part of a flux emerges from the flank of a tooth into the slot at an angle of 1° to the normal. What is the angle which the lines of force make with the side of the slot in the iron, assuming the relative permeability of the iron to be 1000?   Ans. $90° - \theta_i = 3° \ 17'$.

CHAPTER VII

THE MAGNETOMOTIVE FORCE OF DISTRIBUTED WINDINGS

42. The M.M.F. of a Direct-current or Single-phase Distributed Winding. In the two preceding chapters it is shown how to calculate the ampere-turns required for a given flux in an electric machine. When the exciting winding is *concentrated*, that is, when all the turns per pole embrace the whole flux, the number of ampere-turns is equal to the product of the actual amperes flowing through the winding times the number of turns. Such is the case in a transformer, in a direct-current machine, and in a synchronous machine with salient poles. In some cases, however, the exciting windings are *distributed* along the air-gap, so that only a part of the flux is linked with all the turns, and the actual ampere-turns have to be multiplied by a factor in order to obtain the effective m.m.f. Such is the case in an induction motor, and in an alternator with non-salient poles. Moreover, one has to consider the m.m.f. of distributed armature windings when calculating the performance of a machine under load, because the armature currents modify the no-load flux. In this chapter the m.m.fs. of distributed windings are treated mainly in application to the performance of the induction motor; in particular, to the calculation of the no-load current and the reaction of the secondary currents. The armature reaction in synchronous and in direct-current machines is analyzed in the next two chapters.

Distributed Winding for Alternator Field. A cross-section of a four-pole field structure with non-salient poles for a turbo-alternator is shown in Fig. 33a. The flux is graded (Fig. 33b) in spite of a constant air-gap, because the total ampere-turns act only upon the part *a* of a pole; two-thirds of the ampere-turns act upon the parts *b*, *b* and one-third upon the parts *c*, *c*. The m.m.f. and the flux in the parts *d*, *d* are equal to zero. Thus, theoretically, the flux density in the air-gap should vary according to a " stepped "

121

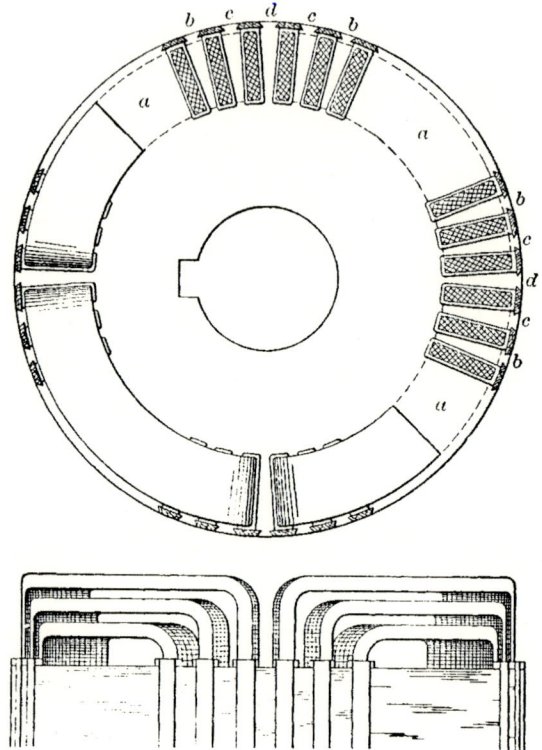

FIG. 33a.   A four-pole revolving fieldstructure with non-salient poles.

curve (Fig. 33b); in reality, the corners are smoothed out by the
fringes.   The total number of ampere-turns per pole must be such

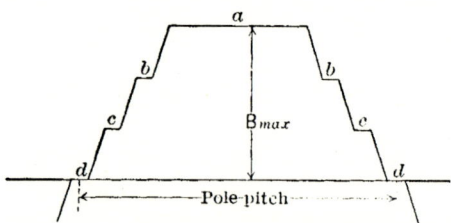

FIG. 33b.   The flux-density distribution for
the field shown in Fig. 33a.

as to create the assumed
maximum flux density
B_{max} in the air-gap
under the middle part
of the pole. The slots
are placed with due re-
gard to the mechanical
strength of the struc-
ture, and so as to get a
flux-density distribu-
tion approaching a sine

wave.   The middle part of the curve is left flat, because very little flux would be gained by placing a narrow coil near the center of the pole, at a considerable expense in copper, and in power loss for excitation.   The total flux, which is proportional to the area of the curve in Fig. 33*b*, must be of the magnitude required by eq. (31) for the induced e.m.f.   If greater accuracy is desired, the curve in Fig. 33*b* is resolved into its fundamental sine wave and higher harmonics; the area of the fundamental curve must then give the flux Φ which enters into eq. (31).

Single-phase Distributed Winding. Let us consider now the stator winding of an induction motor, and in particular the m.m.f. created by the current in one phase.   We begin with the simplest case of a winding placed in one slot per pole per phase (Fig. 34). The reluctances of the stator core and of the rotor core are small as compared with that of the air-gap and the teeth, and are taken into account by increasing the reluctance of the *active layer* of the machine (air-gap and teeth).   If P and Q are the centers of the slots in which opposite sides of a coil are placed, the m.m.f.

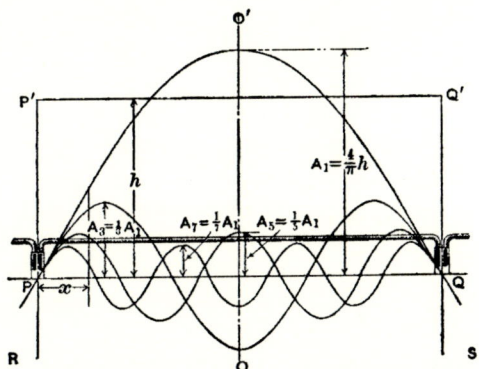

Fig. 34.—The m.m.f. of a single-phase unislot winding resolved into its harmonics.

distribution along the air-gap is that shown by the broken line $RPP'Q'QS$.   In other words, the m.m.f. across the active layer, at any instant, is constant over a pole pitch, and is alternately positive and negative under consecutive poles.

Let n be the number of turns per pole, and i the instantaneous current; then the height PP' of the rectangle is equal to ni. It is understood of course that such an m.m.f. acting alone does not produce the sinusoidal distribution of the flux density assumed in the previous chapters: In a single-phase motor the sinusoidal distribution is due to the simultaneous action of the stator and rotor currents, and also to the fact that the windings are distributed in several slots per pole.   In a polyphase machine the simultaneous action of the two or three phases also helps to secure a sinusoidal distribution. As long as the coil PQ acts alone, the m.m.f. has a " rectangular " distribution in space, and, if the current in the coil varies with the time according to the sine law, the height of the rectangle, or the m.m.f. across the active layer, also varies according to the sine law. In what follows it is important to distinguish between variations of the m.m.fs. in space, i.e., along the air-gap, and those occurring in time, as the current in a winding varies.

For the purposes of analysis the rectangular distribution of the m.m.f. can be replaced by an infinite number of sinusoidal distributions (Fig. 34), according to Fourier's series.[1] The advantages of such a development over the orginal rectangle $PP'Q'Q$ are as follows:

(a) The sine wave is a familiar standard by which all other shapes of periodic curves are judged.

(b) When adding the m.m.fs. due to the coils in different slots, or belonging to different phases, it is much more convenient to add sine waves than to add rectangles displaced in space and varying with the time.

(c) In the actual operation of an induction motor or generator the higher harmonics in the m.m.f. wave are to a considerable extent wiped out by the corresponding currents in the rotor, so that the rectangular distribution is actually changed to a nearly sinusoidal one (see Art. 45 below).

Let h be the height of the rectangle; we assume that for all the points along the air-gap the sum of the ordinates of all the sine waves is equal to h; or

$$h = A_1 \sin x + A_3 \sin 3x + A_5 \sin 5x + \text{etc.} \qquad . \quad . \quad (59)$$

Here x is the angle in electrical degrees, counted along the air-gap, and $A_1, A_3, A_5, \ldots$ are the amplitudes of the waves, to be determined as functions of h. No cosine harmonies enter into this formula, because the m.m.f. distribution is symmetrical with respect to the center line OO' of the exciting coil. To determine the amplitude of the nth harmonic A_n, multiply both sides of eq. (59) by $\sin nx \, d(nx)$, and integrate both sides between the limits $x = 0$ and

---

[1] For the general method of expanding a periodic function into a series of sines and cosines, see the author's *Experimental Electrical Engineering*, Vol. 2, pp. 222 to 227.

$x = \pi$.   All the terms on the right-hand side vanish, except the one containing $\sin^2 nx$, and we have

$$2h = \tfrac{1}{2} n\pi A_n,$$

from which

$$A_n = 4h/(n\pi). \ldots \ldots \ldots (60)$$

Thus, the required series is

$$h = 4h/\pi \ (\sin x + \tfrac{1}{3} \sin 3x + \tfrac{1}{5} \sin 5x + \text{etc.}) \quad . \quad . \quad (61)$$

This means that the amplitude of the fundamental wave is $4/\pi$ times larger than the height h of the original rectangle; the amplitude of the third harmonic is equal to one-third of that of the fundamental wave; the amplitude of the fifth harmonic is one-fifth of that of the fundamental wave, etc.   In practical applications the fundamental wave is usually all we desire to follow, but in some special cases a few of the harmonics are important.[1]

Let now the winding of a phase be distributed in S slots per pole (Figs. 15 and 16), the distance between the adjacent slots being α electrical degrees.   The conductors in every pair of slots distant by a pole pitch produce a rectangular distribution of the m.m.f. like the one shown in Fig. 34, or, what is the same, an equivalent series of sine-wave distributions.   The m.m.fs. produced by the different coils are superimposed, and, since a sum of sine waves having equal bases is also a sine wave, the resultant m.m.f. also consists of a fundamental sine wave and of higher harmonics. The fundamental waves of the m.m.fs. of the several coils are displaced by an angle of α electrical degrees with respect to one another, so that the amplitude of the resultant wave is not quite S times larger than that of each component wave.   The reduction coefficient, or the slot factor, k_s, is the same as that for the induced e.m.f. (Art. 28), because in both cases we have an addition of sine waves displaced by α electrical degrees, (see also prob. 20 in Art.

[1] This method of treating the m.m.fs. of distributed windings by resolving the rectangular curve into its higher harmonics is due to A. Blondel. See his article entitled " Quelques propriétés générales des champs magnétiques tournants," *L'Eclairage Electrique*, Vol. 4 (1895), p. 248.   Some authors consider the actual " stepped " curves of the m.m.f. or flux distribution, a procedure rather cumbersome, and in the end less accurate, in view of the fact that the higher harmonics are to a considerable extent wiped out by the currents in the rotor.

28.) For the same reason, the value of the winding-pitch factor, k_w, deduced in Art. 29, holds for the m.m.fs. as well as for the induced e.m.fs.

When adding the waves of the higher harmonics due to several coils, one must remember that an angle of α electrical degrees for the fundamental wave is equivalent to 3α electrical degrees for the third harmonic, 5α for the fifth harmonic, etc. Therefore, when using the formula (29) and Fig. 19, different values of α and of per cent pitch must be used for each harmonic, and in this connection the reader is advised to review Art. 30. In the practical problems given below the higher harmonics of the armature m.m.f. are disregarded altogether. The results so obtained are in a sufficient agreement with the results of experiments to warrant the great simplification so achieved. For the completeness of the treatment, and as an application of the general method, an analysis of the effect of the higher harmonics of an m.m.f. is given in Art. 45 below. However, this article may be omitted, if desired, without impairing the continuity of the treatment in the rest of the book.

Resolution of a Pulsating m.m.f. into Two Gliding m.m.fs. The reader is aware from elementary study that the pulsating m.m.fs. produced by two or three phases combine into one gliding (revolving) m.m.f. in the air-gap. It is therefore convenient to consider even a single-phase pulsating m.m.f. as a combination of m.m.fs. gliding along the air-gap in opposite directions. In this wise, the m.m.fs. due to different phases are later combined in a simple manner. This method of treatment is similar to that used in mechanics, when an oscillatory motion is resolved into two rotary motions in opposite directions. Also in the analysis of polarized light a similar method of treatment is used.

Take the first harmonic of the m.m.f. (Fig. 34) and assume the current in the exciting coil to vary with the time according to the sine law; then the amplitude of the m.m.f. wave also varies with the time according to the sine law. Imagine two m.m.f. waves, of half the maximum amplitude of the pulsating wave, gliding uniformly along the air-gap in opposite directions; the superposition of these waves gives the original pulsating wave. One can see this by drawing such waves on two pieces of transparent paper and placing them in various positions over a sketch showing the pulsating wave. It will be found that the sum of the corresponding ordinates of the revolving waves gives the ordinate of the pulsating wave at the same point. Or else, represent the two gliding waves by two vectors of equal magnitude M, revolving in opposite directions. The resultant vector is a pulsating one in a constant direction, and varies harmonically between the values $\pm 2M$.

The analytical proof is as follows: Let the exciting current reach its maximum at the moment $t=0$. Then, if the amplitude of the m.m.f. wave at this instant is equal to A, the amplitude at any other instant t is equal to $A \cos 2\pi f t$. Therefore, the m.m.f. corresponding to a point distant x from P and at a time t is equal to $A \cos 2\pi f t \sin x$. By a familiar trigonometrical transformation we have

$$A \sin x \cos 2\pi f t = \tfrac{1}{2} A \sin (x + 2\pi f t) + \tfrac{1}{2} A \sin (x - 2\pi f t). \qquad (62)$$

The right-hand side of this equation represents two sine waves, of the amplitude $\tfrac{1}{2}A$, gliding synchronously along the air-gap, that is, covering one pole pitch during each alternation of the current. The wave $\tfrac{1}{2}A \sin (x + 2\pi f t)$ glides to the left, because, with increasing t, the value of x must be reduced in order to get the same phase of the m.m.f. wave, that is, to keep the value of $(x + 2\pi f t)$ constant. The other wave glides to the right, because, with increasing t, the value of x must be increased in order to obtain any constant value of $(x - 2\pi f t)$. A similar resolution into two gliding waves can be made for each higher harmonic of the pulsating m.m.f. wave; the higher the order of a harmonic the lower the linear speed of its two gliding wave components.

In practice it is usually required to know the relationship between the effective value i of the magnetizing current, the number of turns n per pole per phase, and the crest value of one of the gliding m.m.f. waves. From the preceding explanation this relationship for the fundamental wave is

$$M = \tfrac{1}{2}(4/\pi)k_b n i \sqrt{2} = 0.9 k_b n i, \quad . \quad . \quad . \quad (63)$$

where M is the amplitude of each of the two gliding m.m.fs., $ni\sqrt{2}$ represents the maximum height h of the original rectangle, and the factor $\tfrac{1}{2}$ is introduced because the amplitude of each gliding wave is one-half of that of the corresponding pulsating wave. The breadth factor k_b is the same as that used for the induced e.m.fs. (Arts. 27 to 29). Similar expressions can be written for each higher harmonic, remembering that their amplitudes decrease according to eq. (61),

and that a different value of k_b must be used for each harmonic.
The value of M is calculated so as to produce the required revolving flux, as is explained in Chapters IV, V, and VI. From eq. (63)
either n or i, or their product can be determined.

Prob. 1. A single-phase four-pole induction motor has 24 stator slots,
two-thirds of which are occupied by the winding; there are 18 conductors per slot. The average reluctance of the active layer is 0.09 rel.
per square centimeter. What current is necessary to produce a pulsating
flux of such a value that the maximum flux density due to the first
harmonic is 5 kl./sq.cm., when the secondary circuit is open?

Ans.   8.3 amp.

Prob. 2. Show that in the preceding problem the difference between
the actual flux per pole and its fundamental is less than 2 per cent.

Prob. 3. Show that, if in Fig. 34 the angle x is counted from the
crest of the first harmonic, the expansion into the Fourier series is similar
to eq. (61), except that cosines take place of the sines, and the terms
are alternately positive and negative.

43. The M.M.F. of Polyphase Windings. Consider a two-
phase winding of the stator of an induction motor (Fig. 35a); let

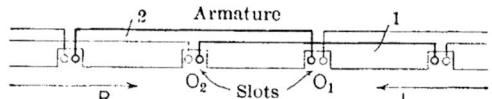

Fig. 35a.—A two-phase winding.

the current in phase 1 lead that in phase 2 by $\frac{1}{4}T$, or by 90 electrical
degrees. A little reflection will show that the resultant m.m.f. of
the two phases glides from right to left: Let the current in phase 1
reach its maximum at the instant $t=0$; at this instant the current
in the coil 2 is zero, and the m.m.f. wave is distributed uniformly
under the coil 1; at the instant $t=\frac{1}{4}T$ the current in phase 1 is zero,
and the m.m.f. is distributed under the coil 2. At intermediate
instants both coils contribute to the resultant m.m.f., so that its
maximum occupies a position intermediate between the centers
O_1 and O_2 of the coils 1 and 2.

The actual rectangular distribution of the m.m.f. due to each
phase can be replaced by a fundamental sinusoidal one and its
higher harmonics, as in Fig. 34. The pulsating fundamental m.m.f.
of each phase can be replaced by two waves of half the amplitude, gliding synchronously in opposite directions. Let the wave

due to phase 1, and gliding to the left, be denoted by L_1, and that due to phase 2 by L_2. Let the corresponding waves gliding to the right be denoted by R_1 and R_2. Disregarding the higher harmonics, the resultant m.m.f. is due to the combined action of the four gliding waves L_1, L_2, R_1 and R_2. At the instant $t=0$ the crest of the wave L_1 is at the point O_1; at the instant $t=\frac{1}{4}T$ the crest of the wave L_2 is at the center O_2 of the coil 2. Consequently, at the instant $t=0$ the crest of the wave L_2 is 90 electrical degrees to the right of O_2, or it is at O_1. Thus, the waves L_1 and L_2 actually coincide in space, and form one wave of double the amplitude.

The crest of the wave R_1 is at the point O_1 when $t=0$; the crest of R_2 is at the point O_2 when $t=\frac{1}{4}T$. Therefore, at $t=0$ the crest of R_2 is 90 electrical degrees to the left of the point O_2, and the waves R_1 and R_2 travel at a distance of 180 electrical degrees from each other. But two such waves cancel each other at all points and at all moments, so that there is no resultant R wave. Thus the resultant fundamental wave of m.m.f. in a two-phase machine is gliding. Its amplitude is twice as large as that of either of the component gliding m.m.fs. of the two phases, which components

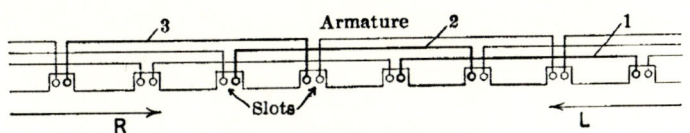

Fig. 35b.—A three-phase winding.

are expressed by eq. (63). If the current in phase 2 were leading with respect to that in phase 1, the L fluxes would cancel each other and the resultant flux would travel from left to right.

Consider now a three-phase winding (Fig. 35b) and call the m.m.fs. which glide to the left, and which are due to the separate phases, by L_1, L_2, and L_3 respectively. Let the waves which travel to the right be denoted by R_1, R_2, and R_3. Assume the current in phase 2 to be lagging by 120 electrical degrees, or by $\frac{1}{3}T$, with respect to that in phase 1, and the current in phase 3 to be lagging by $\frac{1}{3}T$ with respect to that in phase 2. By a reasoning similar to that given for the two-phase winding above it can be shown that the three L waves coincide in their position in space, and give one gliding wave of three times the amplitude of each wave. The three R waves are relatively displaced by 240 electrical degrees, or, what is the same, by 120 electrical degrees; hence, their m.m.fs. mutually cancel at each point along the air-gap. This can be proved by drawing three sine waves displaced by 120 degrees and adding their ordinates point by point; or else one can replace each wave by a vector, and show that the sum of the three vectors is zero because they form an equilateral triangle.

The reasoning given for the two- and three-phase windings can be extended to any number of symmetrical phases, say m, provided that the windings are displaced in space by $360/m$ electrical degrees, and also provided that the currents in these windings are displaced in time by $1/m$th of a cycle. The gliding fundamental waves due to each phase which go in one direction are in phase with each other, and, when added, give a wave m times larger than that expressed by eq. (63); while the fundamental waves going in the opposite direction are displaced in space by $720/m$ electrical degrees, and their combined m.m.f. is zero. The direction in which the resultant m.m.f. travels is from the leading to the lagging phases of the winding. Thus, for any symmetrical m-phase winding

$$M = 0.9k_b mni, \quad \ldots \ldots \ldots \quad (64)$$

where M denotes the amplitude of the fundamental sine wave of the resultant gliding m.m.f., n is the number of turns per pole per phase, and i is the effective value of the current in each phase.

Prob. 4. It is desired to build a 60 horse-power, 550-volt, 4-pole, Y-connected induction motor, using a stator punching with 4 slots per pole per phase, and a winding pitch of one hundred per cent. The required maximum m.m.f. per pole is estimated at 1550 ampere-turns. What is the total required number of stator turns (for all the phases) if the magnetizing current must not exceed 25 per cent of the full-load current? The estimated full-load efficiency is 92 per cent, the power factor at full load is about 90 per cent?   Ans. Not less than 504.

Prob. 5. What is the required number of conductors per slot in the preceding problem, if the stator winding is to be delta-connected and to have a winding pitch of 75 per cent?   Ans. 40.

Prob. 6. What is the amplitude of the first harmonic of the total armature reaction in a 1000 kva, 440 volt, 6 pole, two-phase alternator with non-salient poles? The stator has 72 slots; the coils lie in slots 1 and 9; the number of conductors per slot is C_s. In practice, the armature reaction must not exceed a certain limit, and this helps to determine the permissible value of C_s.   Ans. $4800C_s$ amp.-turns.

Prob. 7. Plot the actual "stepped" curves of m.m.f. distribution for a two-phase winding with three slots per pole per phase, for the following instants: $t = 0$, $t = \frac{1}{16}T$, and $t = \frac{1}{8}T$. Compare the maximum and the average m.m.f. of the actual distribution with those of the first harmonic.[1]

Prob. 8. Solve the preceding problem for a three-phase winding with 2 slots per pole per phase, and a winding pitch of $\frac{5}{6}$. Take two instants, $t = 0$ and $t = \frac{1}{12}T$, and show that for the instants $t = \frac{2}{12}T, \frac{4}{12}T, \frac{6}{12}T$, etc., the m.m.f. distribution is the same as for $t = 0$, while for $\frac{3}{12}T, \frac{5}{12}T$, etc., the m.m.f. distribution is the same as for $t = \frac{1}{12}T$.

Prob. 9. Prove directly that two equal pulsating sine waves of m.m.f. or flux, displaced by 90 electrical degrees in space and in time relatively to each other, give a gliding sine wave, the amplitude of which is equal to that of each pulsating wave. Solution: The left-hand side of eq. (62) gives the value of the m.m.f. at a point x and at an instant t, due to phase 1; the m.m.f. produced at the same point and at the same instant by the phase 2 is $A \sin (x + \frac{1}{2}\pi) \cos (2\pi ft - \frac{1}{2}\pi)$. Adding the two expressions gives $A \sin (x + 2\pi ft)$, which is a left-going wave of amplitude A.

Prob. 10. Prove, as in the preceding problem, that the three pulsating, sine waves of m.m.f. produced by a three-phase winding, give together a gliding m.m.f., the amplitude of which is 50 per cent larger than that of each pulsating wave.

Prob. 11. Prove by the method given in problem 9 above that m pulsating m.m.f. waves displaced in space and in time by an electrical angle $2\pi/m$ produce a gliding m.m.f. the amplitude of which is $\frac{1}{2}m$ times larger than that of each pulsating wave. See Arnold, *Wechselstrom-technik*, Vol. 3 (1908) p. 302.

44. The M.M.Fs. in a Loaded Induction Machine.[2] Eq. (64) gives the magnetizing current i of an induction motor at no-load, i.e., when the rotor is running at practically synchronous speed, so that the secondary currents are negligible. When the motor is loaded, the useful flux which crosses the air-gap is due to the combined action of the primary and the secondary currents. In commercial motors the flux at full load is but a few per cent below that at no load, the difference being due to the impedance drop in the

---

[1] Problems 7 and 8 are intended to acquaint the student with the usual method of calculation of the m.m.fs. of distributed windings and to show the advantage of Blondel's method used in the text. For numerous stepped curves and calculations, see Boy de la Tour, *The Induction Motor*, Chapter IV.

[2] The treatment in this article presupposes a general knowledge of the equivalent performance diagram of induction machines; the purpose of the article being to deduce the exact numerical relations. This article and the one following can be omitted without impairing the continuity of treatment in the rest of the text.

primary winding, the same as in a transformer. Therefore, the net number of exciting ampere-turns, M, is approximately the same as at no load. This means that the geometric sum of the m.m.fs. produced by the primary and the secondary currents at any load is nearly equal to the m.m.f. due to the primary winding alone at no load. In this respect the induction motor is similar to a transformer.

(a) *Calculation of the Secondary Current.* Knowing the primary full-load current, the secondary full-load current can be calculated from the required counter-m.m.f.; the procedure can be best illustrated by an example. In the motor given in prob. 4 above, the full-load current is estimated at 57 amp.; taking the direction of the vector of the applied voltage as the axis of reference, the full-load current can be represented as $51.3 - j24.8$ amp. The magnetizing current, $0.25 \times 57 = 14.25$, is practically in quadrature with the applied voltage, because it is in quadrature with the induced counter e.m.f., the same as in a transformer. The full-load current of 57 amp. contains a component which supplies the iron loss in the stator; we estimate it to be equal to about 1.1 amp. (2 per cent of the input). Thus, the component of the primary current, the action of which must be compensated by the secondary currents, is $(51.3 - j24.8) - (1.1 - j14) = 50.2 - j10.8$ amp., or its absolute value is 51.4 amp. This is called the *current transmitted into the secondary*, or the *secondary current reduced to the primary circuit.* This current produces a maximum m.m.f. of $0.9 \times 3 \times 0.958 \times 42 \times 51.4 = 5580$ amp.-turns.

Let the rotor be provided with a three-phase winding, with 5 slots per pole per phase, and let the winding pitch be 13/15. The number of slots is selected so as to be different from that in the stator, in order to insure a more uniform torque, and to reduce the fluctuations in the reluctance of the active layer. We have, according to eq. (64), that $5580 = 0.9 \times 3 \times 0.935 \times (ni)$, from which $ni = 2210$ amp.-turns. Certain practical considerations, for instance, the value of the induced secondary voltage, usually limit the choice of one of these factors; then the other factor also becomes definite. If, for instance, the rotor is to have 10 conductors per slot, the secondary current will be about 89 amp. The secondary i^2r loss is determined by the desired per cent slip; knowing the secondary current and the number of turns, the necessary size of the conductor can easily be calculated.

Sometimes the secondary winding consists of *coils individually short circuited;* this is an intermediate type of winding between an ordinary squirrel-cage winding and a three-phase winding such as is used with slip-rings. Let the foregoing motor be provided with such a winding, of the two-layer type, and let the rotor have 71 slots, 6 conductors per slot, the coils being placed in slots 1 and 14. In formula (64) m stands for the number of symmetrically distributed phases, the current in each phase being displaced in time by $2\pi/m$ with respect to that in the next phase. In the winding under consideration, each coil represents a phase, and one has to go over a pair of poles until one finds the next coil with the current in the same phase. Thus, in this case, the number of secondary phases is equal to the number of slots per pair of poles, or $m = 35.5$. Each coil has 3 turns, but there is only one coil per pair of poles, so that $n = 1.5$. Substituting these values into eq. (64), and also $M = 5580$, $k_b = 0.912$, we find $i = 128$ amp. As a matter of fact, in this case it is not necessary to decide what the values of n and m are, because eq. (64) contains only the product mn, which is the total number of turns per pole. Thus, in our case $mn = (71 \times 3)/4$.

Formula (64) holds also for a squirrel-cage winding, the number of secondary phases being equal to the number of bars per pair of poles, or $m_2 = C_2/(\tfrac{1}{2}p)$, where C_2 is the total number of rotor bars, and p is the number of poles. Since there is but one bar per phase, each bar can be considered as one-half of a turn, and in formula (64) $n = 0.25$ and $k_b = 1$, so that it becomes

$$M = 0.45iC_2/p. \qquad \ldots \quad (64a)$$

Or else, one may say that the total number of secondary turns per pole is equal to one-half the number of bars per pole, so that $mn = \tfrac{1}{2}C_2/p$. This again gives eq. (64a). For a direct proof of formula (64a) see problem 15 below. Applying this formula to the same rotor with 71 slots we find that the current per bar is 700 amp.

(b) *The Equivalent Secondary Winding Reduced to the Primary Circuit.* When investigating the general theory of the induction motor or calculating the characteristics of a given motor, it is convenient to replace the actual rotor winding by an equivalent winding identical with the primary winding of the motor. In this case the primary current transmitted into the secondary is equal to the actual secondary current (one to one ratio of transformation), and the primary and the secondary voltages induced by the useful flux are also equal. Each electric circuit of the stator then can be combined with the corresponding rotor circuit. In this manner the so-called " equivalent diagram " of the induction motor is obtained,[1] a way of representation which greatly simplifies the theory of the machine.

Let i_2 be the secondary current in the coils or bars of the actual rotor, and i'_2 that in the equivalent rotor. The counter-m.m.f. of both rotors must be the same, this being the condition of their equivalence, so that $0.9k_{b2}m_2n_2i_2 = 0.9k_{b1}m_1n_1i_2'$, from which

$$i_2'/i_2 = (m_2/m_1)(k_{b2}n_2/k_{b1}n_1). \quad . \quad . \quad . \quad (65)$$

This is the ratio of current transformation in an induction motor. The ratio of transformation of the voltages is different, namely,

$$e_2'/e_2 = k_{b1}n_1 \ k_{b2}n_2. \quad . \quad . \quad . \quad . \quad (66)$$

In an ordinary transformer $e_2'/e_2 = i_2/i_2' = n_1/n_2$, because there $k_{b1} = k_{b2} = 1$, and $m_2 = m_1$. For this reason, the induction motor is sometimes regarded as a generalized transformer. For an application to the squirrel-cage rotor see Appendix III.

Taking the product mie for the actual and the equivalent rotor it will be found that the total electric power input is the same in both, provided that the same phase displacement is preserved in the equivalent rotor as in the original one. The latter condition is essential in order that the operating characteristics of the two machines be the same. This means (a) that the total i^2r loss of the equivalent rotor must be equal to that of the original rotor, in order to preserve the same slip, and (b) that the leakage reactances of the two rotors must affect the power factor of the primary current in the same way.

Let r_2 and r_2' be the resistances of the actual and of the equivalent rotor, per pole per phase. We have the condition that

$$m_2i_2^2r_2 = m_1i_2'^2r_2'. \quad . \quad . \quad . \quad . \quad (67)$$

[1] Chas. P. Steinmetz, *Alternating Current Phenomena* (1908), p. 249; *Elements of Electrical Engineering* (1905), p. 263; V. Karapetoff, *The Electric Circuit*, Chapters XII and XIII.

Substituting the ratio of i_2'/i_2 from eq. (65) we find

$$r_2'/r_2 = (m_1/m_2)(k_{b1}n_1/k_{b2}n_2).^2 \quad \cdots \quad (68)$$

For a transformer this equation reduces to the familiar expression $r_2'/r_2 = (n_1/n_2)^2$.[1] For the ratio of resistances in a motor with a squirrel-cage rotor see Appendix III.

The ratio of the inductances is the same as that of the resistances; this can be proved as follows: In order that the equivalent winding may have the same effect on the power factor of the motor as the actual winding, the equivalent winding must draw from the line an equal amount of reactive volt-amperes, due to its leakage inductance.   Equating the magnetic energies stored in the two rotor windings we have, according to eq. (104), Art. 58,

$$m_2 \cdot \tfrac{1}{2}i_2^2 L_2 = m_1 \cdot \tfrac{1}{2}i_2'^2 L_2', \quad \cdots \quad (69)$$

where L_2 and L_2' are the leakage inductances of the real and the equivalent rotor windings, per pole per phase.   The form of this equation is the same as that of eq. (67); therefore, substituting again the ratio of i_2'/i_2 from eq. (65), we have

$$L_2'/L_2 = (m_1/m_2)(k_{b1}n_1/k_{b2}n_2)^2. \quad \cdots \quad (70)$$

This result could also be foreseen from the fact that the reactances and the resistances enter symmetrically in the equivalent diagram, and relation (68) holds therefore for the reactances x_2 and x_2'. But in the equivalent diagram the secondary and the primary frequency is the same, so that the ratio of the inductances is equal to that of the reactances; this gives eq. (70).

It must be clearly understood that the expressions (68) and (70) refer to the resistances and inductances *per pole per phase*. When the windings of a phase are all in series, both in the stator and in the rotor, the same ratio holds of course for the resistances per phase; otherwise the actual connections must be taken into consideration, keeping in mind that the total i^2r loss must be the same in the equivalent winding as in the actual one.   Having obtained the resistance of the equivalent winding per pole, the turns are connected in the same way as the stator turns. This fact must be remembered in particular when dealing with individually short-circuited coils

---

[1] See the author's *Experimental Electrical Engineering*, Vol. 2, p. 77; *The Electric Circuit*, Art. 40.

in the rotor, or with a squirrel-cage winding. In these two cases the individual coils or bars in the rotor are all in parallel, while the stator coils of a phase are usually all in series, or in two parallel groups. In the case of a squirrel-cage winding the resistance of a bar must be augmented by that of two contacts with the end-rings, and of the equivalent resistance of a section of the two end-rings.[1]

Prob. 12. In a 300 horse-power, Y-connected, 14-pole induction motor the full-load current is estimated to be 310 amp. The primary winding consists of 336 turns placed in 168 slots; the winding pitch is 0.75. What is the minimum number of bars in the squirrel-cage secondary winding, if the current per bar must not exceed 800 amp.? The secondary counter-m.m.f. is equal to about 90 per cent of the primary m.m.f.   Ans. 207.

Prob. 13. What must be the resistance of each secondary bar in the preceding problem (including the equivalent resistance of the adjoining segments of the end-rings and also of the contacts) if the slip at full load is to be about 4 per cent.? Hint: The per cent slip is equal to the i^2r loss in the rotor, expressed in per cent of the power input into the secondary. If x is the i^2r loss in the rotor, expressed in horse power, we have that $x = 0.04 \ (300 + x)$.   Ans. 70.4 microhms.

Prob. 14. The motor with the individually short circuited secondary coils, that is used as an illustration in the text above, is to be investigated with respect to its performance. By what factor must the actual resistance and inductance of each secondary coil be multiplied in order to obtain the equivalent resistance and inductance per primary phase? Also by what factor must the equivalent current be multiplied in order to obtain the actual current in each secondary coil?
                                                         Ans. 146; 2.49.

Prob. 15. Prove formula (64a) directly, by considering the m.m.fs. of the individual bars. Solution: At any instant the currents in the bars under a pole are distributed in space according to the sine law, because the gliding flux which induces these currents is sinusoidal. The average current per bar is $i\sqrt{2} \times (2/\pi) = 0.9i$. The number of turns per pole is $C_2/2p$, and all these turns are active at the crest of the m.m.f. wave. Therefore, $M = 0.9i(C_2/2p)$.

45. The Higher Harmonics of the M.M.F's.

In the preceding study, the effect of the higher harmonics in the m.m.f. wave was disregarded. In fact, these harmonics usually exert a negligible influence upon the operation of a good polyphase induction motor, under normal conditions. These m.m.f. harmonics move at lower speeds than the fundamental field; therefore, the fluxes which they

[1] See the author's *Electric Circuit*, Art. 45; also E. Arnold, *Die Wechsel-stromtechnik*, Vol. 5, Part I (1909), p. 57.

produce cut the secondary conductors at comparatively high relative speeds; thus, secondary currents are induced which wipe out these harmonics to a considerable degree. There are practical cases, however, in which some one particular harmonic becomes of some importance, and affects the operation of the machine, particularly at starting. For this reason the following general outline of the properties of the higher harmonics in the m.m.f. is given.[1]

In a single-phase machine (Fig. 34) all the higher harmonics of the m.m.f. are pulsating at the same frequency as the fundamental wave, but the width of the nth harmonic is only $1/n$th of that of the fundamental wave. Each pulsating harmonic can be replaced by two gliding harmonics of half the amplitude, one left-going, the other right-going. The linear velocity of these gliding m.m.fs. is only $1/n$th of that of the fundamental gliding waves, because they cover in the time $\frac{1}{2}T$ a distance equal only to their own base, PQ/n (180 electrical degrees). With one slot per pole, the amplitudes of the higher harmonics decrease according to eq. (61), but with more than one slot, or with a fractional-pitch winding they decrease more rapidly, because different values of k_b must be taken for each harmonic (see Art. 30 above).

In a two-phase machine, consider (Fig. 35a) the gliding waves $\mathbf{L_n}$ and $\mathbf{R_n}$, of the nth harmonic. For this harmonic, the distance between O_1 and O_2 is equal to $\frac{1}{2}\pi n$ electrical degrees. At the instant $t=O$ the crest of the wave $\mathbf{L_{n1}}$ is at the point O_1; at the instant $t=\frac{1}{4}T$ the crest of the wave $\mathbf{L_{n2}}$ is at the point O_2. Therefore, the two waves travel at a relative distance of $\frac{1}{2}\pi(n-1)$ electrical degrees, considering the base of the nth harmonic as equal to its own 180 electrical degrees. In a similar manner, the distance between the crests of the two right-going waves is found to be equal to $\frac{1}{2}\pi(n+1)$ electrical degrees. We thus obtain the following table of the angular distances between the waves due to the two phases:

Order of the harmonic	1	3	5	7	9	11	13
Distance between the two $\mathbf{L_n}$ waves	0	π	2π	3π	4π	5π	6π
Distance between the two $\mathbf{R_n}$ waves	π	2π	3π	4π	5π	6π	7π

The waves which travel at a distance $0, 2\pi, 4\pi$, etc., are simply added together, while those at a distance $\pi, 3\pi, 5\pi$, etc., cancel each

[1] For a more detailed treatment see Arnold, *Wechselstromtechnik*, Vol. 3 (1912), Chapter 10, and Vol. 5, part I (1909), Chapter 9.

other. Thus, in a two-phase machine, the 3d, 7th, 11th, etc., harmonics travel *against the direction* of the main m.m.f., while the 5th, 9th, 13th, etc., harmonics travel *in the same direction* as the fundamental m.m.f., though at lower peripheral speeds.

Applying a similar reasoning to a three-phase winding (Fig. 35b) we find that the three L_n waves travel at a relative distance of $\frac{2}{3}\pi(n-1)$, while the relative distance between the three R_n waves is $\frac{2}{3}\pi(n+1)$ electrical degrees. We thus obtain the following table of the angular distances between the waves due to the three phases:

Order of the harmonic	1	3	5	7	9	11	13	15
Distance between the three L_n waves	0	$\frac{4}{3}\pi$	$\frac{8}{3}\pi$	0	$\frac{16}{3}\pi$	$\frac{20}{3}\pi$	0	$\frac{28}{3}\pi$
Distance between the three R_n waves	$\frac{4}{3}\pi$	$\frac{8}{3}\pi$	0	$\frac{16}{3}\pi$	$\frac{20}{3}\pi$	0	$\frac{28}{3}\pi$	$\frac{32}{3}\pi$

The component waves, of any harmonic, which travel at a distance zero from each other, are simply added together, and give a resultant wave of three times the amplitude of the component. The three waves which travel at an angular distance of $\frac{4}{3}\pi$ or one of its multiples from each other give a sum equal to zero. Thus, in a three-phase machine, the 1st, 7th, 13th, etc., harmonics travel in one direction, while the 5th, 11th, 17th, etc., harmonics travel against the direction of the fundamental m.m.f. The higher the order of a harmonic the lower its peripheral speed. The harmonics of the order 3, 9, 15, etc., are entirely absent.

Prob. 16. What are the amplitudes of the fifth and the seventh harmonics, in percentage of that of the fundamental wave, for a three-phase winding placed in 2 slots per pole per phase, when the winding-pitch is 5/6?   Ans. 1.4 and 1.0 per cent respectively.

Prob. 17. Show that, in order to eliminate the nth harmonic in the m.m.f. wave, the winding-pitch must satisfy this condition; namely, $\gamma/\pi = (2q+1)/n$, where γ is defined in Fig. 16, and q is equal to either 0, 1, 2, 3, etc. Hint: Cos $\frac{1}{2}\gamma n$ must be = 0.

Prob. 18. Investigate the direction of motion of the various harmonics of the m.m.f. in a symmetrical m-phase system.

Prob. 19. Show that only the nth harmonic in the m.m.f. wave, due to the nth harmonic in the exciting current, moves synchronously with the fundamental gliding m.m.f., and therefore distorts it permanently.

Prob. 20. A poorly designed 2-phase, 60-cycle induction motor has 4 poles, 1 slot per phase per pole, and a winding pitch of 100 per cent. At what sub-synchronous speed is it most likely to stick? Hint: The torque due to any harmonic reverses as the motor passes through the corresponding sub-synchronous speed.   Ans. 360 r.p.m.

CHAPTER VIII

ARMATURE REACTION IN SYNCHRONOUS MACHINES

46. Armature Reaction and Armature Reactance in a Synchronous Machine. When a synchronous machine carries a load, either as a generator or as a motor, the armature currents, being sources of m.m.f., modify the flux created by the field coils, and thus influence the performance of the machine. Fig. 36 shows an

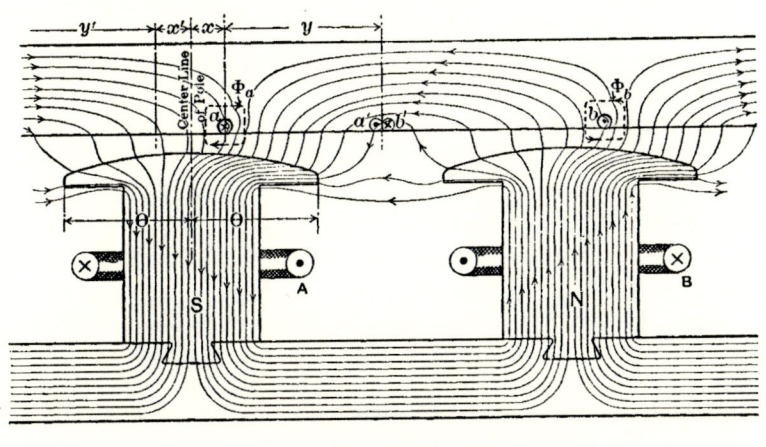

<<----- Direction of Rotation -----<<<<

Fig. 36—The flux distribution in a single-phase synchronous machine under load.

instantaneous flux distribution in the simplest case of a single-phase alternator, with one slot per pole; the armature conductors are marked a and b. With the directions of the armature and field currents indicated in the sketch, the flux is crowded toward the right-hand tips of the poles. In order to show this, imagine two fictitious conductors a' and b' with currents equal and opposite

139

to those in the actual conductors a and b respectively. The addition of these fictitious conductors does not modify the armature m.m.f. because they neutralize each other. The conductor a' may be considered as forming a turn with a, while b' forms a turn with b. It will be seen that the m.m.f. of the coil aa' assists that of the field coil A, while the m.m.f. of the coil bb' is opposite to that of the field coil B.

The armature current in the coil ab not only distorts the no-load field, but also reduces the total flux per pole. This may be seen by considering the flux in the four parts of the air-gap, marked x, y, x', and y', where $x = x'$ and $y = y'$. The sum of the fluxes in the portions y and y' is the same as without the armature current, because the flux density in the part y is increased by the same amount by which it is reduced in the part y' (neglecting saturation). But in the parts x and x' the flux is reduced by the armature m.m.f., so that the total result over the pole-pitch is a reduction in the value of the no-load flux. The position of the armature conductors and the direction of the armature currents have been selected arbitrarily. They can be chosen so that the flux will be crowded toward the left-hand tips of the poles, or so that the total flux will be increased by the armature m.m.f., instead of being reduced. The influence of the armature currents, in modifying the value of the field flux and distorting it, is called the *armature reaction*. The armature reaction is measured in ampere-turns, since it is a magnetomotive force.

In addition to the general distortion of the field by the armature currents, there is a local distortion around each armature conductor. This distortion does not extend into the pole shoes, but is limited to the slots and the air-gap; it is indicated in Fig. 36 by ripples in the flux around a and b. These ripples may be regarded as a result of the superposition upon the main flux of the local fluxes Φ_a and Φ_b excited by the armature currents. While these local fluxes, shown by the dotted lines, have no real existence, except around the end connections of the armature conductors, it is convenient to consider them separately. They are purely alternating fluxes, in phase with the currents with which they are linked, so that they induce in the armature windings alternating e.m.fs. in a lagging phase quadrature with the currents.

The effect of these local fluxes upon the voltage of the machine is represented by a certain *armature reactance*, because the effect is the same as if the armature winding created no leakage fluxes around it, but a separate reactance coil were connected in series with each armature lead. The calculation of the armature reactance, or of the local fluxes, is treated in Art. 67, the subject of this chapter being armature reaction only, that is, the effect of the load upon the main magnetic circuit. In the numerical problems of this chapter, for the solution of which it is necessary to know the value of the armature reactance, this value is given. It is not quite correct, strictly speaking, to separate the local distortion of the main flux as a phenomenon by itself; moreover, the separation is somewhat indefinite and arbitrary. However, the flux so separated is comparatively small, and the treatment of the armature reaction proper is thereby greatly simplified.

The distribution shown in Fig. 36 varies from instant to instant because the relative position of the armature changes with reference to the poles, as well as the value of the armature current. Besides, there are usually two or three armature phases, and several slots per pole per phase. It would be out of the question to calculate the actual fluxes for each instant and to take into account their true influence upon the e.m.f. induced in the armature. In practice, certain approximate average values of armature reaction and of armature reactance are employed, which permit one to predict the actual performance of a machine with a sufficient accuracy.

In the case of a synchronous generator (alternator) the problem usually presents itself in the following form: It is required to predetermine the field ampere-turns necessary for a prescribed terminal voltage at a given load. Knowing the resistance and the leakage reactance of the armature, the voltage drop in the armature is added geometrically to the terminal voltage; this gives the induced voltage in the machine. Knowing from the no-load saturation curve the required *net* excitation at this voltage, and correcting it for the effect of the armature reaction, the necessary field ampere-turns are obtained. The results of such calculations for different values of the armature current and for various power factors, plotted as curves, are called the *load characteristics* of the alternator.

In the case of a synchronous motor the terminal voltage is usually given, and it is required to determine the field excitation such that, at a given mechanical output, the input to the armature be at a given power-factor; a leading power-factor is usually prescribed, in order to raise the lagging power-factor of the whole plant. The problem is solved in like manner to that of the generator, by taking into account the proper signs when calculating the reactance drop and the armature reaction. The results, plotted in the form of curves, are called the *phase characteristics*, or *V-curves* of a synchronous motor.[1]

It will be seen from Fig. 36 that the crowding of the flux to one pole-tip, by the armature currents, is primarily due to the fact that the poles shown there are projecting or salient, so that the reluctance along the air-gap is variable. With non-salient poles the flux is simply shifted sidewise without being distorted. Therefore, before going into the details of the calculation of armature reaction in machines with salient poles we shall first consider (in the next article) the case of a machine with non-salient poles.

Prob. 1. Draw the distribution of the flux, similar to that shown in Fig. 36, when the armature conductors are opposite the centers of the poles, and when they are somewhere between the adjacent pole-tips.

Prob. 2. Explain the details of the flux distribution in Fig. 36, by means of a hydraulic analogy, assuming A and B to represent two main centrifugal pumps, and a and b to be two smaller pumps placed in the stream.

Prob. 3. Let each field coil in Fig. 36 have N turns, and let the exciting current be I; let the number of conductors at a be C_s, and the instantaneous value of the armature current i. What is the total flux per pole, if the average permeance of the machine per pole is $\mathcal{P}$ perms per electrical radian, and the angles θ and x are in electrical radians?

Ans. $(2NI\theta - C_s i x)\mathcal{P}$ in maxwells.

Prob. 4. Let a synchronous machine be loaded in such a way that the armature current reaches its maximum when the conductors a and b (Fig. 36) are opposite the centers of the poles, in other words, the current is in phase with the e.m.f. which would be induced at no load. Prove that (neglecting saturation) the *average flux* per pole during a complete cycle is the same as without the armature reaction, but is crowded to the leading tip of the pole, i.e., in the direction of rotation in the case of a motor, and to the trailing tip, or against the direction of rotation when the machine is working as a generator. Hint: The flux is weakened as much in the position x of the conductors as it is strengthened in the symmetrical position x'; the distortion is in the same direction in both positions.

[1] See the author's "*Experimental Electrical Engineering*," Vol. 2, p. 121; also his "Essays on Synchronous Machinery." *General Electric Review*, **1911**, p. 214.

Prob. 5. Let a synchronous machine be loaded in such a way that the armature current reaches its maximum when the conductors a and b (Fig. 36) are midway between the poles, in other words, when the current is displaced by 90 electrical degrees with respect to the e.m.f. induced at no load. Prove that the *average distortion* during a complete cycle is zero, but that the flux is weakened if the armature current lags behind the induced e.m.f., and is strengthened by a leading current. Hint: The flux is weakened in both of the symmetrical positions, x and x', of the conductors, but the distortion is in opposite directions.

Prob. 6. In a single-phase synchronous machine the armature current reaches its maximum when the armature conductors are displaced by an angle ψ with respect to the centers of the poles;[1] prove that the field is distorted by the component $i \cos \psi$ of the current and is weakened or strengthened by the component $i \sin \psi$.

47. The Performance Diagram of a Synchronous Machine with Non-Salient Poles. Let, in a machine with non-salient poles, the field winding be placed in several slots per pole, so that the field m.m.f. in the active layer of the machine is approximately distributed according to the sine law. Consider the machine to be a polyphase generator supplying a partly inductive load. The amplitude of the first harmonic of the armature reaction has the value given by eq. (64) in Art. 43, and revolves synchronously with the field m.m.f., as is explained there. Since the sum of two sine waves is also a sine wave, the resultant m.m.f. is also distributed in the active layer of the machine according to the sine law.

To deduce the phase displacement, in space, between the two sine waves, consider the coil $a\,b$ (Fig. 36) to be one of the phases of the polyphase armature winding. For reasons of symmetry, the maximum m.m.f. produced by a polyphase winding is at the center of the coil in which at that particular moment the current is at a maximum. Assume first that the current in the phase ab reaches its maximum when the conductors a and b are opposite the centers of the poles. The maximum armature m.m.f. at that instant is displaced by 90 electrical degrees with respect to the center lines of the poles. The direction of the armature current is determined by the well-known rule, and it is found to be such that the armature m.m.f. *lags* behind that of the pole, considering the direction of rotation of the poles as positive. Since both m.m.fs. revolve synchronously, this angle between the two m.m.f. crests is pre-

---

[1] The angle ψ is different from the external phase-angle ϕ between the current and the terminal voltage; see Fig. 37.

served all the time. Thus, in a polyphase generator, the armature m.m.f. lags behind the field m.m.f. by 90 electrical degrees in space, when the currents are in phase with the voltages induced at no-load. This statement is in accord with that in problem 4 in the preceding article, because, if each phase shifts the flux against the direction of rotation, all the phases together simply increase the result.

Let now the currents in the armature windings be lagging 90 electrical degrees behind the corresponding e.m.fs. induced at no-load. This simply means that the armature m.m.f. is shifted further back by 90 degrees as compared to the case considered before; therefore, the angle between the field m.m.f. and the armature m.m.f. is 180 electrical degrees, and the two m.m.fs. are simply in phase opposition. This is in accord with the statement in prob. 5.

From the two preceding cases it follows that, when in a synchronous machine with non-salient poles the currents lag by an angle ϕ electrical degrees (Figs. 37 and 38) with respect to the induced voltage at no-load, the armature m.m.f. wave lags by an angle of $90 + \phi$ electrical degrees behind the field m.m.f. wave. In the case of a generator with leading currents the angle ϕ is negative; in a synchronous motor ϕ is larger than 90 degrees.

Let, in Fig. 37, i be the vector of the current in one of the phases, and let e be the corresponding terminal voltage, the phase angle between the two being ϕ. Adding to e in the usual way the ohmic drop ir in the armature, in phase with i, and the reactive drop ix in leading quadrature with i, the induced voltage E in the same phase is obtained.[1] The resultant useful flux, Φ, which induces this e.m.f. leads E by 90 degrees in time; Φ is in phase with the net or resultant m.m.f. M_n which produces it. The m.m.f. M_n is a sum of the field m.m.f. M_f and of the armature reaction M_a

---

[1] On account of skin effect and eddy currents in the armature conductors, the effective resistance r to alternating currents is considerably higher than that calculated or measured with direct current. The actual amount of increase depends upon the character of the winding, the size of the conductors, the shape of the slots, the frequency, etc., so that no definite rule can be given. Fortunately, the ohmic drop constitutes but a small percentage of the voltage of a machine, so that a considerable error committed in estimating the value of the ir drop affects the voltage relations but very little. See A. B. Field, "Eddy Currents in Large Slot-wound Conductors," *Trans. Amer. Inst. Elect. Engrs.*, Vol. 24 (1905), p. 761.

expressed by eq. (64).   The triangle OFG represents the relations in space, while the figure $OABD$ is a time diagram.   Therefore, the two figures are independent of one another; but it is convenient to combine them into one, by using the common vectors Φ and i.

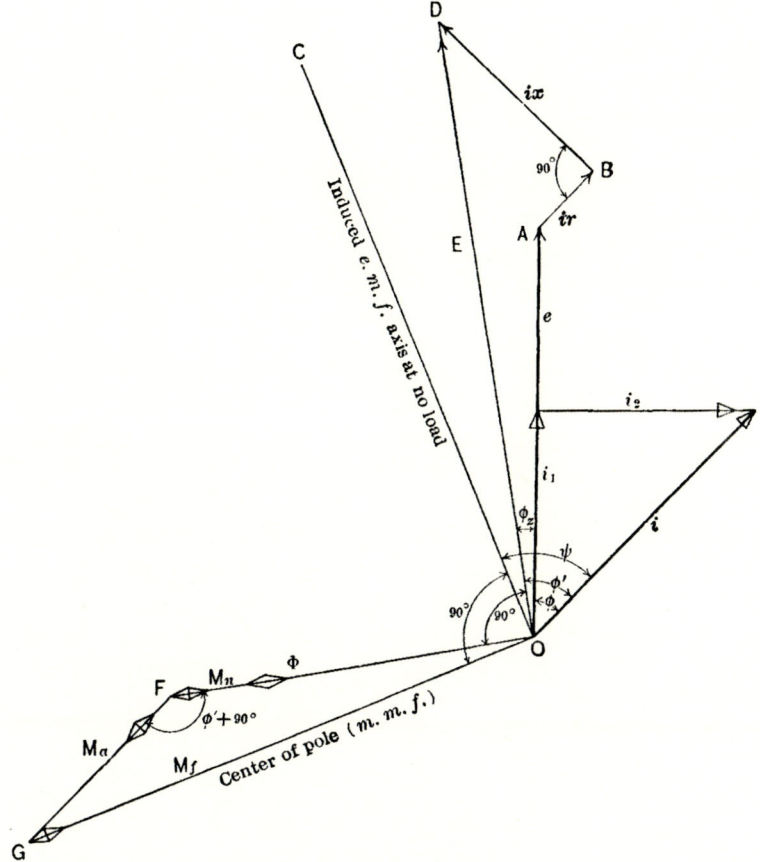

Fig. 37.—The performance diagram of a synchronous *generator*, with *non-salient* poles.

With respect to the triangle OFG, the vector i represents the position of the crest of the armature m.m.f. relatively to the crest OG of the field m.m.f., the angle between the two being $90 + \psi$, as is explained above.   Thus, the vector M_a is in phase with i.

When i and e are given, the vector E is easily found if the resistance and the reactance of the armature winding are known. The required net excitation, M_n, is then taken from the no-load saturation curve of the machine, and M_a is figured out from eq. (64). Then the required field ampere-turns, M_f, are found from the diagram, either graphically or analytically.

The diagram shown in Fig. 37 is known as the *Potier diagram*. Strictly speaking, it is correct only for machines with non-salient poles, but as an approximate semi-empirical method it is sometimes used for machines with projecting poles, in place of the more correct diagram shown in Fig. 40. Fig. 37 represents the conditions in the case of a generator with lagging currents. When the current is leading the vector i is drawn to the left of the vector e, with the corresponding changes in the other vectors.

A similar diagram for a synchronous motor which draws a leading current from the line is shown in Fig. 38. The vector e' represents the line voltage, and e is the equal and opposite voltage which is the terminal voltage of the machine considered as a generator. The rest of the diagram is the same as in Fig. 37. A leading current with respect to the line voltage e' is a lagging current with respect to the generator terminal voltage e, so that the field is weakened by the armature reaction in both cases ($M_n < M_f$ in both figures). The energy component i_1 of the current is reversed in the motor, therefore the field is shifted in the opposite direction; M_n leads M_f in the motor diagram and lags behind it in the generator diagram. The case of a synchronous motor with a lagging current can be easily analyzed by analogy with the above-described cases.

In practice, it is usually preferred to represent the relations shown in Figs. 37 and 38 analytically, rather than to actually construct a diagram. The following relations hold for both the generator and the motor. Projecting all the sides of the polygon $OABD$ on the direction e and on the direction perpendicular to and leading e by 90 degrees, we have

$$E \cos \phi_z = e + ir \cos \phi + ix \sin \phi, \qquad \ldots \quad (71)$$

$$E \sin \phi_z = ix \cos \phi - ir \sin \phi. \qquad \ldots \quad (72)$$

where ϕ_z is the angle between the vectors e and E, counted positive when E leads e, as in Fig. 37. The subscript z suggests that the angle ϕ_z is due to the impedance of the armature.   The expressions $i \cos \phi$ and $i \sin \phi$ represent the energy component and the reactive component of the current respectively; they are designated in

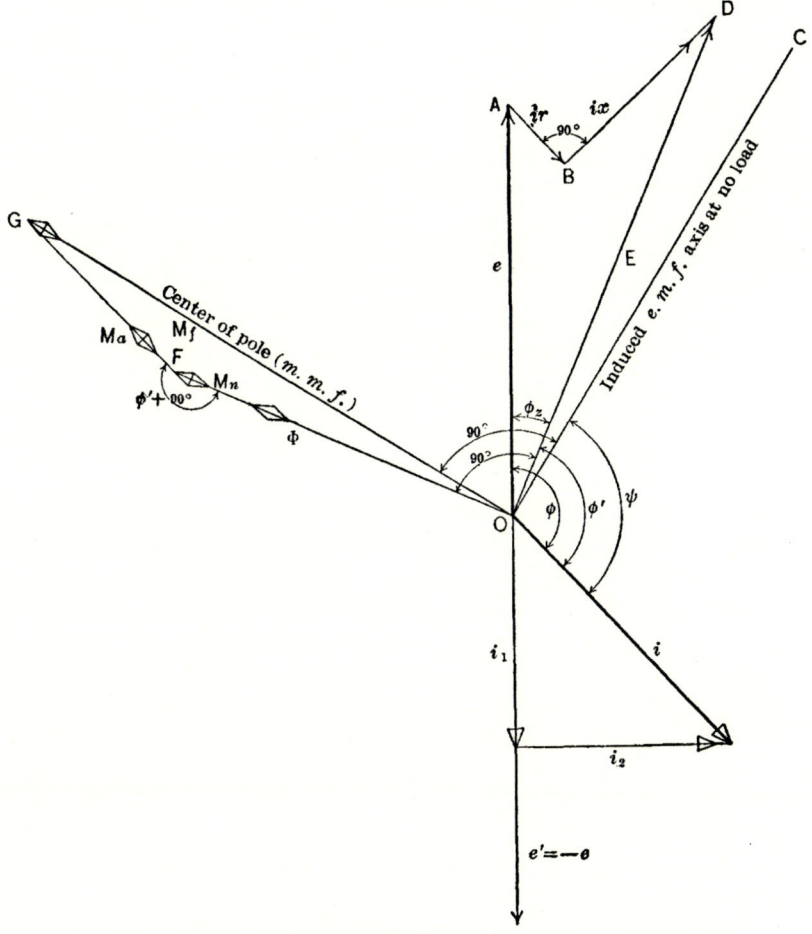

Fig. 38.—The performance diagram of a synchronous *motor*, with *non-salient* poles.

Figs. 37 and 38 by i_1 and i_2.   Denoting the right-hand sides of the eqs. (71) and (72) by e_1 and e_2 for the sake of brevity, we have:

$$e_1 = e + i_1 r + i_2 x; \quad \ldots \ldots \ldots \quad (73)$$

$$e_2 = i_1 x - i_2 r. \quad \ldots \ldots \ldots \quad (74)$$

Squaring eqs. (71) and (72) and adding them together gives

$$E = \sqrt{e_1^2 + e_2^2}.^1 \qquad \ldots \ldots \quad (75)$$

Dividing eq. (72) by (71) results in

$$\tan \phi_z = e_2/e_1. \qquad \ldots \ldots \quad (76)$$

Consequently, the angle between E and i becomes known; namely,

$$\phi' = \phi + \phi_z, \qquad \ldots \ldots \ldots \quad (76a)$$

where ϕ' is called the *internal* phase angle. Knowing E, the corresponding excitation M_n is taken from the no-load saturation curve of the machine; from the triangle OFG we have then:

$$M_f^2 = M_n^2 + M_a^2 + 2M_n M_a \sin \phi', \qquad \ldots \quad (77)$$

where ϕ' is known from eq. (76a). In numerical applications it is convenient to express all the M's in kiloampere-turns.

The diagram shown in Fig. 38 and the equations developed above can be used for determining not only the phase characteristics of a synchronous motor, but its overload capacity at a given field current as well. This latter problem is of extreme importance in the design of synchronous motors. The input into the machine, per phase, is $-ei \cos \phi$; the part ir of the line voltage is lost in the armature, the part ix corresponds to the magnetic energy which is periodically stored in the machine and returned to the line, without performing any work. The remainder, E, corresponds to the useful work done by the machine, plus the iron loss and friction. If the armature possessed no resistance and no leakage reactance the terminal voltage would be equal to E in magnitude and in phase position. Thus, the expression $-Ei \cos \phi'$, corrected for the core loss in the armature iron, represents the input into the revolving structure, per phase. The overload capacity of the machine is determined by the possible maximum of this expression.

The problem is complicated by the fact that the relation between E and M_n is expressed by the no-load saturation curve, which is difficult to represent by an equation. The problem is

---

[1] In numerical applications it is more convenient to use the approximate formula

$$E = e_1 + \tfrac{1}{2}e_2^2/e_1 \quad \ldots \ldots \ldots \ldots \quad (75a)$$

obtained by the binomial expansion of expression (75); since all other terms can be neglected when e_2 is small as compared to e_1.

therefore solved by trials, assuming a certain reasonable value of E, and calculating the expression $-Ei \cos \phi'$, until a value of E is found, for which this expression, corrected for the core loss, friction, and windage, becomes a maximum. The problem of finding i and ϕ' for an assumed E is a definite one, because the four equations (71), (72), (76a) and (77) contain only four unknown quantities, i, ϕ, ϕ_z, and ϕ'. Instead of solving the problem by trials, an analytical relation can be assumed between E and M_n, on the useful part of the no-load saturation curve, for instance a straight line (not passing through the origin), a parabola, etc. The problem is then solved by equating the first derivative of the product—$Ei \cos \phi'$ to zero, having previously expressed E, i, and $\cos \phi'$ through some one independent variable. Both methods have been worked out for a synchronous motor with salient poles.[1] The relations are simplified for a machine with non-salient poles.

The foregoing theory of the armature reaction does not apply directly to single-phase machines. The pulsating armature reaction in such a machine can be resolved into two revolving reactions, as in Art. 42. The reaction which revolves in the same direction with the main field is taken into account as in a polyphase machine. The inverse reaction is partly wiped out by the eddy currents produced in the metal parts of the revolving structure; it is therefore difficult to express the effect of this reaction theoretically. The treatment in this book is limited to polyphase machines, which are used in practice almost exclusively.[2]

Prob. 7. In the 1000 kva., 440-volt, 6-pole, two-phase alternator, given in Problem 6, Art. 43, the amplitude of the first harmonic of the armature reaction was $4800 C_s$ ampere-turns. What is the per cent voltage regulation of the machine at a power-factor of 80 per cent lagging, if $C_s = 1$, that is if the armature has one conductor per slot? The armature reactance is 0.038 ohm, and the armature resistance is 0.008 ohm, both per phase. The no-load saturation curve of the machine is as follows:

$E = 400 \quad 440 \quad 490 \quad 525 \quad 550$ volts.
$M_n = 6.7 \quad 8.0 \quad 10.0 \quad 12.0 \quad 14.0$ kiloamp.-turns.

Ans. 22 per cent.

[1] See the author's "Essays on Synchronous Machinery," *General Electric Review*, 1911, July and September.

[2] In regard to the armature reaction in single-phase machines, see E. Arnold, *Die Wechselstromtechnik*, Vol. 4 (1904), pp. 32–39; Pichelmayer, *Dynamobau* (1908), pp. 251–259; Max Wengner, *Theoretische und Experimentelle Untersuchungen an der Synchronen Einphasen-Maschine* (Oldenbourg, 1911.)

Prob. 8. The machine specified above is to be used as a synchronous motor. Determine graphically the required field excitation when the useful output on the shaft is to be 700 kw., and in addition the machine must draw from the line 600 leading reactive kva. The efficiency of the machine at the above-mentioned load is estimated to be about 91 per cent.   Ans. 12.5 kiloampere-turns.

Prob. 9. Draw to the same scale as the diagram shown in Fig. 37, another similar diagram, for the same value of the current and of the phase angle ϕ, except that the current is to be leading. Assume a reasonable shape of the saturation curve in determining the new value of M_n. Show that a much smaller exciting current is required with the same kva. output, than in the case of a lagging current.

Prob. 10. Solve problem 9 for the motor diagram shown in Fig. 38, assuming the current to be lagging with respect to the line voltage.

Prob. 11. For a given alternator, show how to determine the voltage e (Fig. 37), analytically or graphically, when M_f, i, and ϕ are given; explain when such a case arises in practice.

Prob. 12. For a given synchronous motor, show how to determine the reactive component i_2 of the current (Fig. 38), analytically or graphically, when M_f, e and i_1 are given; explain when such a case arises in practice.

Prob. 13. Work out the details of the above-mentioned method for the determination of the overload capacity of a synchronous motor by trials. Hint: Introduce the components of e and i, in phase and in quadrature with E; rewrite eqs. (71) and (72) by projecting the figure $OABD$ on the direction of E and on that perpendicular to E. Use no angles in the formulæ, and neglect the small terms containing r, where they lead to complicated equations of higher degrees.

48. The Direct and Transverse Armature Reaction in a Synchronous Machine with Salient Poles.

In a machine with non-salient poles the armature reaction shifts the field flux but hardly distorts its shape. In a machine with projecting poles the flux, generally speaking, is both altered in value and crowded toward one pole-tip (Fig. 36). It is convenient, therefore, to resolve the traveling wave of the armature m.m.f. into two waves, one whose crests coincide with the center lines of the poles, the other displaced by 90 electrical degrees with respect to it. The first component of the armature m.m.f. produces only a " direct " effect upon the field flux, that is, it either strengthens or weakens the flux, without distorting it. The second component produces a " transverse " action only, viz., it shifts the flux toward one or the other pole-tip, without altering its value (that is, neglecting the saturation).

We have seen before that an armature current, which reaches its maximum when the conductor is opposite the center of the pole, distorts the flux; while a current in quadrature with the former exerts a direct reaction only. It is natural, therefore, to resolve the actual current in each phase into two components, in time quadrature with each other, and in such a way that each component reaches its maximum in one of the above-mentioned principal positions of the conductor with respect to the field-poles. Let the current in each phase be i, and let it reach its maximum at an angle ψ after the induced no-load voltage is a maximum (Fig. 40). Then, the two components of the current are

$$i_d = i \sin \psi$$

and

$$i_t = i \cos \psi.$$

The component i_d produces a direct armature reaction only, and the component i_t a transverse reaction only.[1]

For practical calculations, and in order to get a concrete picture of the armature reaction, it is convenient to represent the armature reaction as shown in Fig. 39. Namely, the direct reaction, due to the components i_d of the armature currents, is replaced by an equivalent number of concentrated ampere-turns M_d on the pole. The value of M_d is selected so that its action in reducing or strengthening the flux is equal to the true action of the armature currents. The transverse reaction, due to the component i_t of the armature currents, is replaced by a certain number of ampere-turns, M_t, on the fictitious poles, (S), (N), shown by dotted lines between the real poles. For simplicity, and for other reasons given in Art. 51, the fictitious poles are assumed to be of a shape identical with that of the real poles. The number of exciting ampere-turns M_t is so chosen, that the effect of the fictitious poles is approximately the same as that of the distorting ampere-turns on the armature.

The flux of the fictitious poles strengthens the flux of the real poles on one side and weakens it by the same amount on the other side, so that the fictitious poles actually distort the main flux without altering its value. Strictly speaking, the complete action of the distorting ampere-turns on the armature cannot be imitated

---

[1] The resolution of the armature reaction in a synchronous machine into a direct and a transverse reaction was first done by A. Blondel. See *l'Industrie Électrique*, 1899, p. 481; also his book *Moteurs Synchrones* (1900), and two papers of his in the *Trans. Intern. Electr. Congress*, St. Louis, 1904, Vol. 1, pp. 620 and 635.

by fictitious poles of the same shape as the main poles, because harmonics of appreciable magnitude are thereby neglected. However, actual experience shows that the performance of a machine, calculated in this way, can be made to check very well with the observed performance, by properly selecting the coefficients of the direct and the transverse reaction. In a generator, the flux is crowded against the direction of rotation of the poles (Fig. 36); consequently, the fictitious poles lag behind the real poles, as shown in Fig. 39. In a synchronous motor they lead the real poles by 90 electrical degrees.

If the ratio of the pole arc to pole-pitch were equal to unity, as with non-salient poles, the whole wave of the demagnetizing

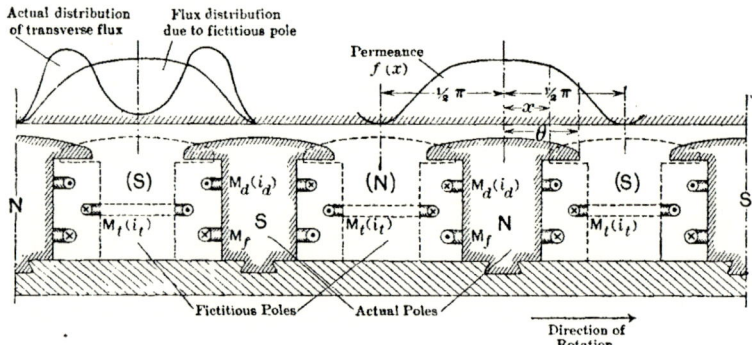

Fig. 39.—The direct and transverse armature reactions in a synchronous machine, represented by fictitious poles and field windings.

m.m.f. of the armature would be acting upon the pole, and the equivalent concentrated m.m.f. M_d on the pole would have to be equal to the average value of the actual distributed armature m.m.f. We would have then

$$M_d = (2/\pi)M \sin \phi, \quad \ldots \quad \ldots \quad (78)$$

where the maximum armature m.m.f. is determined by eq. (64), Art. 43, and $2/\pi = 0.637$ is the ratio of the average to the maximum ordinate of a sine wave. In reality, only a part of the armature m.m.f., the one near its amplitude, acts upon the poles, the action of lower parts of the wave being practically zero because of the gaps between the poles. Therefore, the ratio between the maximum m.m.f. $M \sin \phi$ and the average equivalent m.m.f. M_d is larger than 0.637. For the ordinary shapes of projecting poles, experiment and calculation (see Art. 50 below) show that this ratio varies between 0.81 and 0.85. Using an average of these limits instead of $2/\pi$ in eq. (78) and substituting for M its expression from eq. (64) we obtain the following practical formula for estimating the armature demagnetizing ampere-turns per pole in a synchronous machine with projecting poles:

$$M_d = 0.75 k_b mni \sin \psi. \quad \ldots \ldots \quad (79)$$

In this formula $i \sin \psi$ is the component i_d of the armature current, per phase. In actual machines the numerical coefficient in this formula varies between 0.73 and 0.77, depending on the shape of the poles and the ratio of pole-arc to pole-pitch.

By a similar reasoning, if the ratio of pole-arc to pole-pitch were equal to unity, the equivalent number of exciting ampere-turns on the fictitious poles would be

$$M_t = (2/\pi) M \cos \psi. \quad \ldots \ldots \quad (80)$$

Since the ratio of pole-arc to pole-pitch on the fictitious poles is less than unity, the numerical coefficient should be larger than $2/\pi$. But, on the other hand, the permeance of the air-gap under the fictitious poles is much higher than the actual permeance of the machine in the gaps between the poles, so that a much smaller number of ampere-turns M_t is sufficient to produce the same distorting flux. The combined effect of these two factors is to reduce the coefficient in formula (80) to a value considerably below $2/\pi$. For the usual shapes of projecting poles, experiment and calculation (See Art. 51 below) show that this ratio varies between 0.30 and 0.36. Using an average of these limits instead of $2/\pi$ in eq. (80), and substituting for M its expression from eq. (64), we obtain the following practical formula for estimating the distorting ampere-turns per pole, in a synchronous machine with projecting poles:

$$M_t = 0.30 k_b mni \cos \psi. \quad \ldots \ldots \quad (81)$$

In this formula $i \cos \psi$ is the component i_t of the armature current, per phase. In some actually built machines the coefficient in this formula comes out lower than 0.30, but in preliminary calculations it is advisable to use at least 0.30. When a synchronous motor is working near the limit of its overload capacity, the influence of the distorting ampere-turns is particularly important, and in estimating the overload capacity of a synchronous motor it is better to be on the safe side and to take the value of the numerical coefficient in eq. (81) somewhat higher than 0.30. The value of this coefficient varies within wider limits than that of the corresponding coefficient in formula (79); but, fortunately, it affects the performance to a lesser degree (see Art. 51).

49. The Blondel Performance Diagram of a Synchronous Machine with Salient Poles. Having replaced the actual armature reaction by two m.m.fs. M_d and M_t (Fig. 39) the electromagnetic relations in the machine become those indicated in Figs. 40 and 41. Fig. 40 refers to a generator and is analogous to Fig. 37; Fig. 41 refers to a motor and is analogous to Fig.

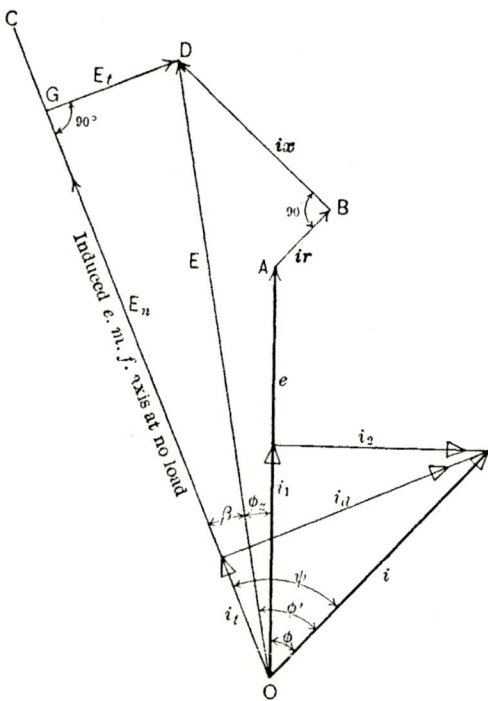

FIG. 40.—The performance diagram of a synchronous *generator*, with *salient* poles.

38. The polygon $OABD$, which represents the relation between the terminal and the induced voltages, is the same as before, but the induced voltage E is now considered as a resultant of the voltages E_n and E_t induced by the real and the fictitious poles respectively.[1] In the generator the fictitious poles lag behind

---

[1] The subscript n stands for net, to agree with the m.m.f. M_n used later on; the subscript t stands for transverse.

the real ones, in a motor they lead the real poles. Hence, in the generator diagram, E_t lags 90 degrees behind E_n, while in the motor diagram it leads E_n by 90 degrees.

In the case of a generator the problem usually is to find the field excitation M_f neces-sary for maintaining a required terminal voltage e, with a given current i and at a given power-factor $\cos \phi$. First, the figure $OABD$ is con-structed, or else the values of E and ϕ' are determined from eqs. (75), (76), and (76a). In order to find the ampere-turns required on the main poles it is neces-sary to determine the voltage E_n induced by them. For this purpose the angle β must first be known, for

$$E_n = E \cos \beta . . \quad (82)$$

As an intermediate step, it is necessary to express E_t through the ampere-turns M_t, which are the cause of E_t. The m.m.f. M_t is small as compared to the total number of ampere-turns on the real poles; hence, the lower straight part of the no-load saturation curve of the machine can be used

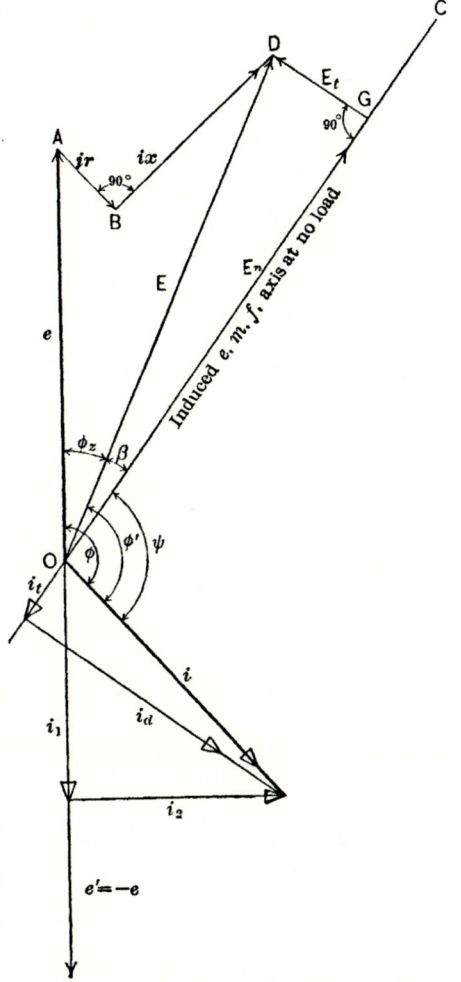

FIG. 41.—The performance diagram of a syn-chronous *motor*, with *salient* poles.

to express the relation between M_t and E_t. Let v be the voltage corresponding to one ampere-turn on the lower part of the no-load saturation curve; then $E_t = M_t v$. Substituting the value of M_t from eq. (81), we have

$$E_t = E_t' \cos (\phi' + \beta), \quad \ldots \ldots \quad (83)$$

where

$$E_t' = 0.30 k_b m n i v. \quad \ldots \ldots \ldots \quad (84)$$

E_t' is a known quantity introduced for the sake of brevity. The angle ϕ in formula (83) is expressed through ϕ' and β, because, from Fig. 40,

$$\psi = \phi' + \beta. \quad \ldots \ldots \ldots \quad (85)$$

Another relation between E_t and β is obtained from the triangle ODG, from which

$$E_t = E \sin \beta. \quad \ldots \ldots \ldots \quad (86)$$

A comparison of eqs. (83) and (86) gives that

$$E_t' \cos (\phi' + \beta) = E \sin \beta.$$

Expanding and dividing throughout by $\cos \beta$ we find the relation

$$\tan \beta = \frac{\cos \phi'}{(E/E_t') + \sin \phi'}, \quad \ldots \ldots \quad (87)$$

from which the angle β can be determined, and then E_n calculated by eq. (82).

The next step is to take from the no-load saturation curve the value M_n of the net excitation necessary on the main poles in order to induce the voltage E_n. The real excitation M_f must be larger, because part of it is neutralized by the direct armature raction M_d. We thus have

$$M_f = M_n + M_d, \quad \ldots \ldots \ldots \quad (88)$$

where M_d is calculated from eq. (79), the angle ϕ being known from eq. (85). When the load is thrown off, the only excitation left is M_f; let it correspond to a voltage e_o on the no-load saturation curve. From e_o and e the per cent voltage regulation of the machine is determined from its definition as the ratio $(e_o - e)/e$.

The same general method and the same equations apply in the case of Fig. 41, when one is required to determine a point on one of the phase characteristics of a synchronous motor. The beginner must be careful with the sign minus in the case of the motor.

Since $\phi' > 90$ degrees, the angle β and the voltage E_t are negative. The angle ϕ_z also is usually negative. The cases of a leading current in the generator and of a lagging current in the motor are obtained by assigning the proper value and sign to the angle ϕ. For the application of the Blondel diagram to the determination of the overload capacity of a synchronous motor see the reference given near the end of Art. 47.

A synchronous motor is sometimes operated at no load, and at such a value of the field current that the machine draws reactive leading kilovolt-amperes from the line, thus improving the power-factor of the system. In such a case the machine is called a *synchronous condenser*, or better, a *phase adjuster*. The diagram in Fig. 41 is greatly simplified in this case because the energy component of the current can be neglected, as well as the drop ir, and the e.m.f. E_t. We then have $i = i_2 = i_d$, and $E_n = E = e + ix$. The direct armature reaction is determined from eq. (79) in which $\psi = 90$. When the motor is underexcited and draws a lagging current from the line, i is to be considered negative, or $\psi = 270$ degrees. The same simplified diagram applies to a polyphase rotary converter, operated from the alternating-current side, at no load.

Prob. 14. It is required to calculate the field current and per cent voltage regulation of a 12-pole, 150 kva., 2300-volt, 60-cycle, Y-connected alternator, at a power factor of 85 per cent lagging. The machine has two slots per pole per phase, and is provided with a full-pitch winding, the number of turns per pole per phase being 18. The armature resistance per phase of Y is 0.67 ohm, the reactance is 3.5 ohm. The number of field turns per pole is 200. The no-load saturation curve is plotted for the line voltage (not the phase voltage), and at first is a straight line such that at 1800 volts the field current is 17.4 amp. The working part of the no-load saturation curve is as follows:

Kilovolts	2.2	2.4	2.5	2.6	2.7	2.78
Field current, amp.	22	25	27	30	34	40

Ans. 31 amp.; 14.3 per cent.

Prob. 15. Show that in the foregoing machine the short-circuit current is equal to about two and a half times the rated current, at the field excitation which gives the rated voltage at no-load. Hint: The short-circuit curve is a straight line so that one can first calculate the field current for any assumed value of the armature current and $e = 0$.

Prob. 16. From the results of the calculations of the preceding problem show that the cross-magnetizing effect and the ohmic drop are negligible under short-circuit, in the machine under consideration.

Assuming that r is usually small as compared to x, describe a simple method for calculating the short-circuit curve, using only the reactance of the machine and the demagnetizing ampere-turns of the armature. In practice, the influence of the neglected factors is accounted for in short-circuit calculations by taking $\sin \psi$ in formula (79) as equal to between 0.95 and 0.98 instead of unity.

Prob. 17. Plot the no-load phase characteristic of the machine specified in problem 14, when it is used as a motor. The iron loss and friction amount to 8.5 kw.

Ans. Field amperes........ 14.9   23.4   32.6
Armature amperes.... 30   2.13   30

Prob. 18. The machine specified in problem 14 is to be used as a motor, at a constant input of 150 kw. Plot its phase characteristics, i.e., the curves of the armature current and of power-factor against the field current as abscissæ.

Ans. Field amperes 32.6   24.3   16.65
Armature amperes..   47.00   37.65   47.00
Power-factor.......   0.80   1.00   0.80

Prob. 19. Write complete instructions for the predetermination of the regulation of alternators and of the phase characteristics of synchronous motors, by Blondel's method. The instructions must give only the successive steps in the calculations, without any theory or explanations. Write directions and formulæ on the left-hand side of the sheet, and a numerical illustration on the right-hand side opposite it.

Prob. 20. Calculate the overload capacities of the foregoing motor at field currents of 25 amp. and 35 amp., by the two methods described in the articles refered to near the end of Art. 47.

Prob. 21. Show that for a machine with non-salient poles Blondel's and Potier's diagrams are identical.

50. The Calculation of the Value of the Coefficient of Direct Reaction in Eq. (79).[1] The average value 0.83 of the ratio of the effective armature m.m.f. over a pole-face to the maximum m.m.f. at the center of the pole is given in Art. 48 without proof. The following computations show the reasonable theoretical limits of this ratio. If the armature m.m.f. (direct reaction) at the center of the N pole (Fig. 39) is M, its value at some other point along the air-gap is $M \cos x$, where x is measured in electrical radians. Let the permeance of the active layer of the machine per electrical radian be $\mathcal{P}$ at the center of the pole, and let this permeance vary along the periphery of the armature according to a law $f(x)$, so that at a point determined by the abscissa x the permeance per

[1] This and the next article can be omitted, if desired, without impairing the continuity of treatment.

electrical radian is $\mathcal{P}f(x)$.   The function $f(x)$ must be periodic and such that $f(0) = 1$, and $f(\tfrac{1}{2}\pi) = 0$, $f(\pi) = 1$, etc., because the permeance reaches its maximum value under the centers of the poles and is practically *nil* midway between the poles.

The direct armature m.m.f., acting alone, without any excitation on the poles, would produce in each half of a pole a flux

$$\Phi = \int_0^{+\frac{1}{2}\pi} M \cos x \, \mathcal{P}f(x)dx.$$

The magnetomotive force M_d placed on the real poles, acting alone, must produce the same total flux, so that

$$\Phi = M_d \int_0^{+\frac{1}{2}\pi} \mathcal{P}f(x)dx.$$

Equating the two preceding expressions we get

$$M \int_0^{\frac{1}{2}\pi} \cos x f(x)dx = M_d \int_0^{\frac{1}{2}\pi} f(x)dx. \quad \ldots \quad (89)$$

The ratio of M_d to M can be calculated from this equation, by assuming a proper law $f(x)$ according to which the permeance of the active layer varies with x, in poles of the usual shapes.   Having a drawing of the armature and of a pole, the magnetic field can be mapped out by the judgment of the eye, assisted if necessary by Lehmann's method (Art. 41 above).   A curve can then be plotted, giving the relative permeances per unit peripheral length, against x as abscissæ.   Thus, the function $f(x)$ is given graphically, and the two integrals which enter into eq. (89) can be determined graphically or be calculated by Simpson's Rule.   Or else, $f(x)$ can be expanded into a Fourier series and the integration performed analytically.   Such calculations performed on poles of the usual proportions give values of M_d/M of between 0.81 and 0.85.

It is also possible to assume for $f(x)$ a few simple analytical expressions, and integrate eq. (89) directly.   Take for instance $f(x) = \cos^2 x$.   By plotting this function against x as abscissæ the reader will see that the function becomes zero midway between the poles, is equal to unity opposite the centers of the poles, and has a reasonable general shape at intermediate points.   Substituting $\cos^2 x$ for $f(x)$ into eq. (89) and integrating, gives $\tfrac{2}{3}M = \tfrac{1}{4}\pi M_d$, from which $M_d/M = 0.85$.

Another extreme assumption is that of poles without chamfer, with a constant air-gap. Neglecting the fringe at the pole-tips, $f(x) = 1$ from $x = 0$ to $x = \theta$, and $f(x) = 0$ from $x = \theta$ to $x = \frac{1}{2}\pi$. Integrating eq. (89) between the limits 0 and θ we obtain

$$M_d/M = (\sin \theta)/\theta. \quad . \quad . \quad . \quad . \quad . \quad (90)$$

The poles usually cover between 60 and 70 per cent of the periphery. For $\theta = 0.6(\frac{1}{2}\pi)$ the preceding equation gives $M_d/M = 0.86$, and for $\theta = 0.7(\frac{1}{2}\pi)$, $M_d/M = 0.81$.

Prob. 22. Let the permeance of the active layer decrease from the center of the poles according to the straight-line law, so that

$$f(x) = 1 - (2/\pi).x.$$

What is the ratio of M_d/M?   Ans. 0.811.

Prob. 23. The permeance of the active layer decreases according to a parabolic law, that is, as the square of the distance from the center of the poles. What is the ratio of M_d/M?   Ans. 0.774.

Prob. 24. The law $f(x) = \cos^2 x$ assumed in the text above presupposes that the permeance varies according to a sine law of double frequency with a constant term, because $\cos^2 x = \frac{1}{2} + \frac{1}{2} \cos 2x$. In reality, the permeance varies more slowly under the poles and more rapidly between the poles than this law presupposes (Fig. 39). A correction can be brought in by adding another harmonic of twice the frequency to the foregoing expression, thus making it unsymmetrical, and of the form $f(x) = a + b \cos 2x + c \cos 4x$. Show that $f(x) = 2 \cos^2 x - \cos^4 x$ contains the largest relative amount of the fourth harmonic, consistent with the physical conditions of the problem, and compare graphically this curve with $f(x) = \cos^2 x$.

Prob. 25. What is the value of M_d/M for the form of $f(x)$ given in the preceding problem?   Ans. 0.815.

Prob. 26. Plot the curve $f(x)$ for a given machine, estimating the permeances by Lehmann's method, and determine the value of the coefficient in formula (79).

51. The Calculation of the Value of the Coefficient of Transverse Reaction in Eq. (81). The average value 0.33 of the ratio of the maximum distorting armature m.m.f. to the equivalent number of ampere-turns, M_t, on the fictitious poles is given in Art. 48 without proof. The following computations show the reasonable theoretical limits of this ratio. The problem is more complicated than that of finding the ratio of M_d/M, because there the field ampere-turns, the actual demagnetizing armature-m.m.f., and the equivalent ampere-turns M_d are all acting on the same permeance of the active layer, and the wave form of the flux is very little affected by the direct armature reaction. In the case of the transverse reaction, however, the wave form of the flux produced by the actual cross-magnetizing ampere-turns of the armature is entirely different from that produced by the coil M_t acting on the fictitious pole (Fig. 39). Namely, the actual curve of the transverse flux has a large " saddle " in the middle, due to the large reluctance of the space between the real poles. The flux distribution produced by the fictitious poles is practically the same as that under the main poles, the two sets of poles being of the same shape.

The addition of the vectors E_t and E_n in Figs. 40 and 41 is legitimate only when E_t is induced by a flux of the same density distribution as E_n, and this is the reason for representing the transverse reaction as due to fictitious poles of the same shape as the real poles. Therefore, for the purposes of computation, the flux distribution, produced by the actual distorting ampere-turns on the armature, is resolved into a distribution of the same form as that produced by the main poles and into higher harmonics. The m.m.f. M_t is calculated so as to produce the first distribution only. This fundamental curve is not sinusoidal, but will have a shape depending on the shape of the pole shoes. The effect of the sinusoidal higher harmonics on the value of E_t is disregarded, or it can be taken into account by correcting the value of the coefficient in formula (81) from the results of tests.

The first harmonic of the armature distortion m.m.f. is $M \sin x$, because this m.m.f. reaches its maximum between the real poles; x is measured as before from the centers of the real poles. The permeance of the active layer, with reference to the real poles, can be represented as before by $\mathcal{P}f(x)$. The flux density produced by the transverse reaction of the armature at a point defined by the abscissa x is therefore proportional to $M \sin x\, \mathcal{P}f(x)$. The permeance of the active layer with reference to the fictitious poles is $\mathcal{P}f(x+\tfrac{1}{2}\pi)$. The flux density under the fictitious poles follows therefore the law $M_t\mathcal{P}f(x+\tfrac{1}{2}\pi)$. As is explained before, the two distributions of the flux density differ widely from one another, and the real distribution is resolved into the fictitious distribution, and higher sinusoidal harmonics; the prominent third harmonic is clearly seen in Fig. 39. Thus, we have, omitting $\mathcal{P}$,

$$M \sin x\, f(x) = M_t f(x+\tfrac{1}{2}\pi) + A_3 \sin 3x + A_5 \sin 5x + \text{etc.} \quad (91)$$

In order to determine M_t, the usual method is to mulitply both sides of this equation by sin x and integrate between 0 and π, because then all upper harmonics give terms equal to zero. In this particular case the limits of integration can be narrowed down to 0 and $\frac{1}{2}\pi$, because the symmetry of the curve is such that the segment between 0 and $\frac{1}{2}\pi$ is similar to all the rest. Thus, we get

$$M \int_0^{\frac{1}{2}\pi} \sin^2 x f(x) dx = M_t \int_0^{\frac{1}{2}\pi} \sin x f(x + \tfrac{1}{2}\pi) dx. \quad . \quad (92)$$

From this equation the ratio M_t/M can be calculated by the methods shown in Art. 50, i.e., by assuming reasonable forms of the function $f(x)$. Taking again $f(x) = \cos^2 x$ and integrating eq. (92) we get $\frac{1}{16}\pi M = \frac{2}{3}M_t$, from which $M_t/M = 0.295$. Taking the other extreme case, viz., $f(x) = 1$ from $x = 0$ to $x = \theta$, and $f(x) = 0$ from $x = \theta$ to $x = \frac{1}{2}\pi$, gives, after integration

$$M_t/M = \{\tfrac{1}{2}\theta - \tfrac{1}{4}\sin 2\theta\}/\sin\theta. \quad . \quad . \quad . \quad . \quad (93)$$

For $\theta = 0.6(\frac{1}{2}\pi)$, $M_t/M = 0.29$; for $\theta = 0.7(\frac{1}{2}\pi)$, $M_t/M = 0.39$.[1] It will be noted that the cross-magnetizing action of the armature increases considerably with the increasing ratio of pole-arc to pole-pitch, while the direct reaction slowly diminishes with the increase of this ratio. In machines intended primarily for lighting purposes it is advisable to use a rather small ratio of pole-arc to pole-pitch, in order to reduce transverse reaction which affects the voltage regulation at high values of power-factor in particular.

Prob. 27. What is the value of M_t/M for the form of $f(x)$ given in problem 24; namely, for $f(x) = 2\cos^2 x - \cos^4 x$?   Ans. 0.368.

Prob. 28. Determine the numerical value of the coefficient in formula (81) for the machine used in problem 26.

---

[1] These values are higher than those derived by E. Arnold. The fact that Arnold's values for the coefficient of transversal reaction are low has been pointed out by Sumec in *Elektrotechnik und Maschinenbau*, 1906, p. 67; also by J. A. Schouten, in his article " Ueber den Spannungsabfall mehrphasiger synchroner Maschinen, "*Elektrotechnische Zeitschrift*, Vol. 31 (1910), p. 877.

CHAPTER IX

ARMATURE REACTION IN DIRECT-CURRENT MACHINES

52. The Direct and Transverse Armature Reactions. Let Fig. 42 represent the developed cross-section of a part of a direct-current machine, either a generator or a motor. For the sake of simplicity the brushes are shown making contact directly with the armature conductors, omitting the commutator. Electrically this is equivalent to the actual conditions, because the commutator segments are soldered to the end-connections of the same conductors. The brushes are shifted by a distance δ from the geometrical neutral, to insure a satisfactory commutation; δ being expressed in centimeters, measured along the armature periphery, the same as the pole-pitch τ.

The actual armature conductors and currents are replaced, for each pole-pitch, by a *current sheet*, or *belt*, of the same strength. Let, for instance, the pole-pitch be 40 cm., and let the machine have 120 armature conductors per pole. If the current per conductor is 100 amp., the total number of ampere-conductors per pole is 12,000; the total current of the equivalent belt, which consists of one wide conductor, must be 12,000 amp., or 300 amp. per linear cm. of the pole-pitch. The latter value, or the number of armature ampere-conductors per centimeter of periphery, is sometimes called the *specific electric loading* of the machine. The magnetic action of the equivalent current sheet on the magnetic flux of the machine is practically the same as that of the actual armature conductors, because in a direct-current machine the slots are comparatively numerous and small. The current in the cross-hatched belts is supposed to flow from the reader into the paper, and the current in the belts marked with dots—toward the reader. With the directions of the flux and of the current shown in the figure, the directions of rotation of the machine when working as a generator and as a motor are those shown by the arrow-heads (see Art. 24).

163

The polarity of the brushes cannot be indicated without knowing the actual connections in the winding. It is preferable, therefore, for our purposes to designate the brushes as E and W (east and west), according to their position with respect to the poles of the machine, the observer looking from the commutator side. The whole interpolar regions to the right of the north poles can be called the eastern regions, those to the left the western regions; the same notation can be also applied to the commutating poles.

The armature currents exert a two-fold action upon the main field of the machine: they partly distort it, and partly weaken it. For the purposes of theory and calculation it is convenient to separate these two actions, the same as in the case of the synchro-

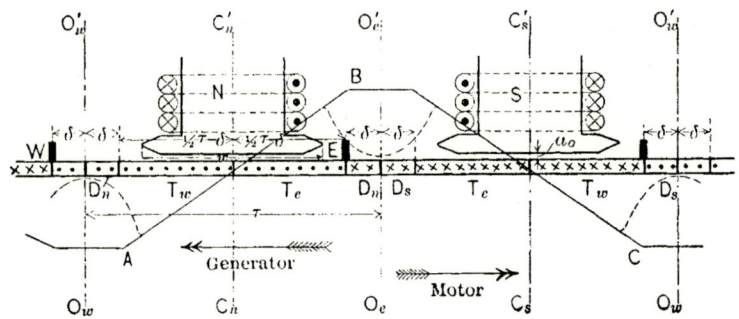

FIG. 42.—The direct and transverse armature reactions in a direct-current machine.

nous machine in the preceding chapter. Let the sheets of current be divided into parts denoted by the letters D and T with subscripts corresponding to their location with reference to the poles and brushes. The belts denoted by D are comprised within the space δ, on each side of the geometrical neutrals; those denoted by T are $(\frac{1}{2}\tau - \delta)$ centimeters wide.

The belts D exert a direct *demagnetizing* action upon the poles. Namely, the belts D_n D_n can be considered as two sides of a coil the axis of which is along the center line $C_n C_n'$. The m.m.f. of this coil opposes that of the field coil on the north pole. In the same way, the m.m.f. of the coil $D_s D_s$ opposes the action of the field coil on the south pole. The foregoing is true no matter what the actual connections of the armature conductors are, provided that the winding-pitch is nearly 100 per cent. With a fractional-pitch winding the currents within each D belt flow partly in the opposite directions and neutralize each other's action.

Let the specific electric loading of the machine, as defined above, be (AC). Then, with a full-pitch winding, the demagnetizing ampere-turns per pole are[1]

$$M_1 = (AC)\partial. \qquad \qquad (94)$$

The belts T_eT_e constitute together a coil the center of which is along the axis O_eO_e'; the adjacent belts T_wT_w form a coil with its axis along O_wO_w'. The m.m.f. distribution of these coils is indicated by the broken line ABC, which shows that the T belts produce a *transverse* armature reaction. The line ABC is also the curve of the flux density distribution which would be produced by the transversal reaction alone, if the active layer of the machine were the same throughout (non-salient poles). On account of a much higher reluctance of the paths in the interpolar regions the flux density there is much lower, and is shown by the dotted lines. The actual distribution of the field in a loaded machine is obtained considering from point to point the field and armature m.m.fs. acting upon the individual magnetic paths.

The transverse reaction opposes the field m.m.f. under one-half of each pole and assists it under the other half, so that the main field is distorted. In a generator the field is shifted in the direction of rotation, in a motor it is crowded against the direction of rotation of the armature. This is the same as what happens in synchronous machines, when the armature is revolving and the poles are stationary (see Fig. 36).

The brushes must be shifted in the same direction in which the flux is shifted, because the magnetic neutral is displaced with respect to the geometric neutral. Usually, the brushes are shifted beyond the magnetic neutral, in order to obtain a proper flux density for commutation. Namely, to assist the reversal of the current in the conductors which are short-circuited by the brushes, these conductors must be brought into the fringe of a field of such a direction as assists the commutation. In the case of a generator this means the field under the influence of which the conduc-

---

[1] The effect of the coils short-circuited under the brushes is not considered separately, for the sake of simplicity. For an analysis of the reaction of the short-circuited coils upon the field see E. Arnold, *Die Gleichstrommaschine*, Vol. 1 (1906), Chap. 23.

tors come *after* the commutation. In a motor the armature current flows against the induced e.m.f.; it is therefore the field which cuts the conductors *before* the commutation that assists the reversal of the current. This explains the direction of the shift of the brushes in the two cases. The student should make this clear to himself by considering in detail the directions of the currents and of the induced voltages in a particular case.

The maximum m.m.f. per pole produced by the distorting belts is equal to (AC) $(\frac{1}{2}\tau - \delta)$, but since this m.m.f. acts along the interpolar space of high reluctance its effect is not large (except in machines with commutating poles). Of much more importance is the action of the distorting belts under the main poles. At each pole-tip the armature m.m.f. is

$$M_2 = (AC)\cdot\tfrac{1}{2}w, \qquad \ldots \ldots (95)$$

where w is the width of the pole shoe. This m.m.f. decreases according to the straight line law to the center of each pole and is of opposite signs at the two tips of the same pole.

Prob. 1. Determine the polarity of the brushes in Fig. 42 for a progressive and a retrogressive winding, in the case of a generator and of a motor.

Prob. 2. Indicate the D and the T belts in a fractional-pitch winding (a) with the brushes in the geometric neutral, and (b) with the brushes shifted by δ.

Prob. 3. A 500 kw., 230 volt, 10 pole, direct-current machine has a full-pitch multiple winding placed in 165 slots. There are 8 conductors per slot, and two turns per commutator segment. What are the demagnetizing ampere-turns per pole when the brushes are shifted by 4 commutator segments? Ans. 3478.

Prob. 4. What is the amplitude of the broken line ABC in the preceding machine? Ans. 10860 amp-turns.

Prob. 5. For a given machine, draw the curves of the flux density distribution under a pole, at no-load and at full load, by considering the m.m.fs. acting upon the individual paths, and the reluctance of the paths.[1]

53. The Calculation of the Field Ampere-turns in a Direct-current Machine under Load. The *net* ampere-turns per pole are determined from the no-load saturation curve of the machine for

[1] For details and examples of such curves see Pichelmayer, *Dynamobau* (1908), p. 176; Parshall and Hobart, *Electric Machine Design* (1906), p. 159; Arnold, *Die Gleichstrommachine*, Vol. 1 (1906), p. 324.

the necessary induced voltage. In a generator the induced voltage is equal to the specified terminal voltage plus the internal ir drop in the machine. In a motor the induced voltage is less than the line voltage by the amount of the internal voltage drop. The actual ampere-turns to be provided on the field poles are larger than the net excitation by the amount necessary for the compensation of the armature reaction.

The direct reaction is compensated for by adding to each field coil the ampere-turns given by eq. (94). For instance, in prob. 3 above, 3470 ampere-turns per pole must be added to the required net excitation, in order to compensate for the effect of the direct armature reaction. The necessary shift of the brushes is only roughly estimated from one's experience with previously built machines, though it could be determined more accurately from the distribution of flux density in the pole-tip fringe. The poles usually cover not over 70 per cent of the armature periphery, so that the distance between the geometric neutral and the pole-tip is about 15 per cent of the pole pitch. In preliminary estimates, the brush shift may be taken to be about 10 per cent of the pole pitch; this brings the brushes not quite to the pole-tips though well within their fringe. In actual operation a smaller shift may be expected. In machines provided with commutating poles, and in motors intended for rotation in both directions, the brushes are usually in the geometric neutral, so that the demagnetizing action is zero.

In a machine with a low saturation in the teeth and in the pole-tips, the cross-magnetizing m.m.f. of the armature does not affect the magnitude of the total flux per pole, because the flux is increased on one-half of the pole as much as it is reduced on the other half. It is shown in Art. 31 that the induced e.m.f. of a direct-current machine is independent of the flux distribution, provided that the total flux is the same, so that no extra field ampere-turns are necessary in such a machine to compensate for the distortion of the flux.

However, the teeth and the pole-tips are usually saturated to a considerable extent, so that the flux is increased on one side of the pole less than it is reduced on the other side. Thus, the useful flux is not only distorted by the transverse armature reaction, but is also weakened. This latter effect has to be counterbalanced by additional ampere-turns on the field poles. In most cases these additional ampere-turns are estimated empirically, on the basis of one's previous experience, because the amount is not large, and a correct computation is rather tedious.[1]

The theoretical relation between the distorting ampere-turns and the field ampere-turns required for their compensation is shown in Fig. 43. Let OX represent the no-load saturation curve of the machine for its active layer only, that is, for the air-gap, the teeth, and the pole shoe, if the latter is sufficiently saturated. The ordinates represent the induced e.m.f. between the brushes, or, to another scale, the useful flux per pole; the abscissæ give the corresponding values of the field ampere-turns per pole, disregarding the reluctance of the field poles, field yoke, and armature core.

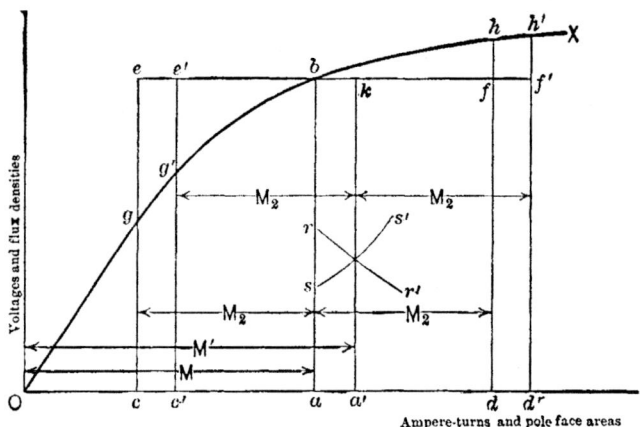

Fig. 43.—A construction for determining the field m.m.f. needed for the compensation of the transverse reaction.

In other words, the abscissæ give the values of the difference of magnetic potential across the active layer of the machine, at no load. It is proper to consider here the m.m.fs. across the active layer only, because the distorting action of the armature extends only over this layer. No matter how irregular the flux distribution in the air-gap and in the teeth may be, the flux density in the pole cores and in the yoke is practically uniform (compare Fig. 36).

[1] For Hobart's empirical curves for estimating the field excitation required for overcoming the armature distortion see the *Standard Handbook*, under 'Generators, direct-current, ampere-turns, estimate."

For the sake of simplicity, we replace the actual pole shoe by an equivalent one, without chamfer, of rectangular shape, and with a negligible pole-tip fringing. The ordinates of the curve OX represent to another scale the average flux density on the surface of the pole shoe, because this density is proportional to the total useful flux, or to the induced voltage. Let ab be the flux density corresponding to the required induced e.m.f., and let $Oa = M$ be the necessary net m.m.f., without the transverse reaction. Let ac and ad represent the distorting ampere-turns, M_2, at the pole-tips, as given by eq. (95); then Oc and Od are the resultant m.m.fs. at the pole-tips. The flux density at the pole-tips of the loaded machine is then equal to cg and dh respectively.

Since the distorting ampere-turns vary directly as the distance from the center of the pole, and the area of strips of equal width is the same, the abscissæ from a represent to scale either distances along the pole face, or areas on the pole face, measured from the center of the pole. Thus, the part gh of the curve OX represents also the distribution of the flux density under the pole, in the loaded machine. Likewise, the line ef represents the distribution of the flux density under the pole at no-load.

Since the ordinates represent to scale the flux densities and the abscissæ the areas of the different parts under the poles, the area under the flux density distribution curve also represents to scale the total flux per pole. The total flux at no load is represented by the area of the reactangle $cefd$, and to the same scale, the flux in the loaded machine is represented by the area $cghd$. If the saturation curve were a straight line, the two areas would be equal, so that the distortion would not modify the value of the total flux per pole. In reality, the area geb is larger than the area bhf, and the difference between the two represents the reduction in the flux, due to the transverse armature reaction.

Let now the field excitation be increased by an unknown amount aa' to the value $Oa' = M'$, in order to compensate for the above explained decrease in the flux. All the points in Fig. 43 are shifted by the same amount, and are denoted by the same letters with the sign " prime." The new flux in the loaded machine is represented by the area $c'g'h'd'$, the point a' being the center of the line $c'd' = cd$. If the point a' has been properly selected we must have the condition that the

$$\text{area } cefd = \text{area } c'g'h'd', \quad . \quad . \quad . \quad . \quad (96)$$

that is, the flux corresponding to the excitation Oa at no-load is the same as the flux corresponding to the excitation Oa' when the machine is loaded. The problem is then, knowing the point a and the distance $cd = c'd'$ to find the point a' such that equation (96) is satisfied. The two areas have the common part $c'g'bfd$, and the parts $cee'c'$ and $dff'd'$ are equal to each other. Therefore, eq. (96) is satisfied if the

$$\text{area } be'g' = \text{area } bf'h'. \quad \ldots \ldots \quad (97)$$

The position of the point a' is found either by trials, or as the intersection of the curves rr' and ss'; the ordinates of the curve rr' represent the area $be'g'$ for various assumed positions of the point a', and the ordinates of the curve ss' represent the corresponding values of the area $bf'h'$.

Thus, the total field ampere-turns required for a direct-current generator under load are found as follows: (a) To the specified terminal voltage add the voltage drop in the armature, under the brushes, and in the windings (if any) which are in series with the armature, viz.; the series field winding, the compensating winding, and the interpole winding. This will give the induced voltage E. (b) From the no-load saturation curve find the excitation corresponding to E. (c) Estimate the amount of the brush shift (if any) and calculate the corresponding demagnetizing ampere-turns according to eq. (94). (d) Determine the ampere-turns aa' (Fig. 43) required for the compensation of the armature distortion. (e) Add the ampere-turns calculated under (b), (c) and (d). In the case of a motor subtract the voltage drop in the machine from the terminal voltage, to find the induced e.m.f. E, but otherwise proceed as before.

If the machine is shunt-wound or series-wound, the field winding is designed so as to provide the necessary maximum number of ampere-turns at a required margin in the field rheostat for the specified voltage or speed variations. When the machine is compound wound, the shunt winding alone must supply the required number of ampere-turns at no-load. The series ampere-turns is, then, the difference between the total m.m.f. required at full-load and that supplied by the shunt-field. When a generator is over-compounded, the shunt excitation is larger at full load than it is at no load, and allowance must be made for this fact.

In the case of a variable-speed motor, the problem of predicting the exact speed at a given load can be solved by successive approximations only. First, this speed is determined neglecting the transverse armature reaction altogether, or assigning to it a reasonable value. Then, the construction shown in Fig. 43 is performed for a few speeds near the approximate value, and thus a more correct value of the speed is found by trials. Or else different values of speed are assumed first, and the corresponding values of the armature current are found for each speed. These armature currents must be such that, considering the total armature reaction, the field m.m.f. is just sufficient to produce the required counter-e.m.f. in the armature. The details of the solution are left to the student to investigate. He must clearly understand that the problem can be solved only for a given or an assumed speed, because of the necessity of using the active layer characteristic OX (Fig. 43).

Prob. 6. In a direct-current machine it is desired to have at full load a flux density of not less than 3500 maxwells per sq.cm. at the pole-tip at which the commutation takes place; the m.m.f. across the active layer is 7500 amp.-turns, the air-gap is 11 mm., the air-gap factor 1.25. The ratio of the ideal pole width to the pole pitch is 0.7. What are the permissible ampere-turns on the armature, per pole? Solution: The net m.m.f. across the active layers at the pole-tip under consideration is $3500 \times 0.8 \times 1.1 \times 1.25 = 3850$ amp.-turns. Hence $M_2 = \frac{1}{2}(AC) \times 0.7\tau = 7500 - 3850 = 3650.$ Ans. $\frac{1}{2}(AC)\tau = 5200.$

Prob. 7. The specific electric loading in a direct-current machine is 250 amp.-conductors per centimeter; the average flux density on the pole-face is 8.5 kl. per sq. cm., at no load and at the rated full load terminal voltage; the width of the equivalent ideal pole is 32 cm. The estimated internal voltage drop at full load is about 5 per cent of the terminal voltage. Calculate the ampere-turns per pole required to compensate for the transverse armature reaction; the active layer characteristic (Fig. 43) is as follows:

$M =$ 4 5 6 7 8 9 11 13 kilo ampere-turns;
$B =$ 5.40 6.75 7.75 8.50 8.90 9.20 9.55 9.78 kl. per sq. cm.

Ans. 950.

Prob. 8. For the preceding machine, calculate the total required excitation at full load, per pole; the brush shift is 8 per cent of the pole-pitch; the pole-pitch $\tau = 46$ cm.; 2000 ampere-turns are required for the parts of the machine outside the active layer.

Ans. 12,000 amp.-turns.

Prob. 9. A shunt-wound motor is designed so as to operate at a certain speed at full load. Show how to predict its speed at no-load.

Prob. 10. Show how, in a compound-wound motor, the total required field excitation must be divided between the shunt and series windings in order to obtain a prescribed speed regulation between no-load and full load.

Prob. 11. Assume the curve OX in Fig. 43 to be given in the form of an analytic equation, $B = f(M)$. Show that the unknown excitation $Oa' = M'$ is determined by the equation $2M_2 f(M) = F(M' + M_2) - F(M' - M_2)$, where the function F is such that $dF(M)/dM = f(M)$; $M_0 = Oa$ is the excitation corresponding to the given value of e or $B = ab$.

Prob. 12. Apply the formula of the preceding problem to the case when the active layer characteristic can be represented by (a) the logarithmic curve $y = a \log (1 + bx)$; (b) the hyperbola $y = gx/(h + x)$; (c) a part of the parabola $(y^2 - y_0^2) = 2p(x - x_0)$, continued as a tangent straight line passing through the origin.

54. Commutating Poles and Compensating Windings.

The two limiting factors in proportioning a direct-current machine are, first, the sparking under the brushes, and secondly, the armature reaction. In order to reverse a considerable armature current in a coil during the short interval of time that the coil is under a brush, an external field of a proper direction and magnitude is necessary. In ordinary machines (Fig. 42) this field is obtained by shifting the brushes so as to bring the short-circuited armature conductors under a pole fringe. However, with this method the specific electric loading and the armature ampere-turns must be kept below a certain limit with reference to the ampere-turns on the field; otherwise the armature reaction would weaken the field to such an extent as to reduce the flux density in the fringe below the required value. Therefore, in many modern machines, instead of moving the brushes to the poles, part of the poles, so to say, are brought to the brushes (Fig. 44). These additional poles are called *commutating poles* or *interpoles*. Their polarity is understood with reference to Fig. 42: Since the E brush had to be shifted toward the north pole, now a north interpole is placed over each E brush.

The armature belts T_c, T_c, create a strong m.m.f. along the axis of the commutating pole N_c, in the wrong direction. Therefore, the winding on each interpole must be provided first with a number of ampere-turns equal and opposite to that of the armature, and secondly with enough additional ampere-turns to establish the required commutating flux. These additional ampere-turns are calculated only for the air-gap, armature teeth, and the pole body itself. The m.m.f. required for the armature core and the yoke of the machine is negligible, because the commutating flux is small as compared with the main flux and is displaced with respect to it by ninety electrical degrees. The winding on the interpoles is connected in series with the main circuit of the machine, because the armature m.m.f. is proportional to the armature current, and also because the density of the reversing field must be proportional to the current undergoing commutation.

The flux density under the interpoles is determined from the condition that the e.m.f. induced in the armature conductors by the commutating flux be equal and opposite to the e.m.f. due to the inductance of the armature coils undergoing commutation. For practical purposes, the inductance can be only roughly estimated (see Art. 68 below), but on the other hand an accurate cal-

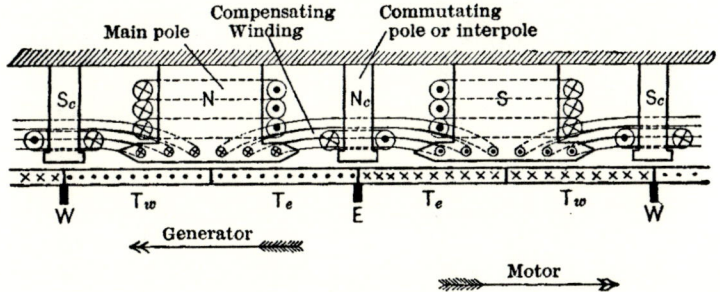

Fig. 44.—Interpoles and a compensating winding in a direct-current machine.

culation is not necessary, because the number of ampere-turns on the interpole is easily adjusted by a shunt around its winding, as in the case of a series winding on the main poles. Or else the commutating flux can be increased by " shimming up " the interpoles. The m.m.f. required for establishing a required commutating flux is calculated in the same manner as in the case of the main flux, viz., the saturation curves (Figs. 2 and 3) are used for the pole body and the teeth, while the reluctance of the air-gap is estimated as is explained in Arts. 36 and 37. The commutating poles must be of such a width that all the coils undergoing commutation are under their influence.

A comparatively large leakage factor, say between 1.4 and 1.5, or over, is usually assumed for the commutating poles, on account of the proximity of the main poles. It is advisable to concentrate the winding near the tip of the interpole, in order to reduce the magnetic leakage. The leakage factor of the main poles is also somewhat increased by the presence of the interpoles; this is one of their disadvantages. Some other disadvantages are: the ventilation of the field coils is more difficult, and a smaller ratio of pole arc to pole pitch must be used. However, the advantages gained by the use of commutating poles are such that their use is rapidly becoming universal.

The interpole winding removes the effect of the transverse belts T, T, in the commutating zone, but does not neutralize their distorting effect under the main poles. Hence, the distortion and its accompanying reduction of the main flux are practically the same as without the interpoles. To remove this distortion a *compensating winding* (Fig. 44), connected in series with the main circuit of the machine, is sometimes placed on the main poles. The connections are such that the compensating winding opposes the magnetizing action of the armature winding. By properly selecting the specific electric loading of the compensating winding the transverse armature reaction under the poles can be removed, either completely or in part. This winding was invented independently by Déri in Europe and by Professor H. J. Ryan in this country; on account of its expense, it is used in rare cases only.

When a compensating winding is used in addition to the interpoles, the number of ampere-turns on the interpoles is considerably reduced, because the compensating winding can be made to neutralize the larger portion or all of the armature reaction. In such a machine a much higher specific loading can be allowed than in an ordinary machine of the same dimensions. Therefore, such compensated machines are particularly well adapted for rapidly fluctuating loads, and for sudden overloads or reversals of rotation in the case of a motor.

Prob. 13. From the following data determine the ampere-turns required on each commutating pole of a turbo-generator: The commutating poles are made of cast steel; the average flux density on the face of the interpole is 6000 maxwells per sq.cm.; the pole-face area 250 sq.cm.; the pole cross-section 160 sq.cm.; the radial length of the interpole 27 cm.; the leakage factor, 1.5. The air-gap reluctance is 2.7 millirels, the true tooth density 20 kilolines per sq.cm., the height of the tooth 4.5 cm., and the armature ampere-turns per pole 9500.

Ans. About 15,300.

Prob. 14. The rated current of the machine in the preceding problem is 1200 amp., and in addition to the interpoles the machine is to be provided with a compensating winding. Show that each interpole should have at least 8 turns, and each main pole be provided with 10 bars for the compensating winding, in order to neutralize the armature distortion under the main poles and provide the proper commutating field. Assume a ratio of pole arc to pole pitch of about 60 per cent.

Prob. 15. Machines provided with interpoles are very sensitive as to their brush position. By shifting the brushes even slightly from the geometrical neutral the terminal voltage of such a generator can be varied to a considerable extent. In the case of a motor the speed can be regulated by this method, without adjusting the field rheostat. Give an explanation of this " compounding" effect of the brush shift.

55. Armature Reaction in a Rotary Converter. The actual currents of irregular form which flow in the armature winding of a rotary converter may be considered as the resultants of the direct current taken from the machine and of the alternating currents taken in by the machine. The resultant magnetizing action upon the field is the same as if these two kinds of currents were flowing through two separate windings. Therefore, the armature reaction in a rotary converter can be calculated by properly combining the armature reactions of a synchronous motor and of a direct-current generator.

The alternating-current input into a rotary converter may be either at a power factor of unity, if the field excitation is properly adjusted, or the input may have a lagging or a leading component, the same as in the case of a synchronous motor. The armature reaction due to the energy component of the input consists chiefly in the distortion of the field, *against the direction of rotation* of the armature. But the action of the direct current is to distort the field *in the direction of rotation,* and since the two m.m.fs. are not much different from one another, the resultant transverse armature reaction is very small. The direct reaction of the direct current depends upon the position of the brushes, and the direct reaction due to the alternating currents is determined by the reactive component of the input, which component may vary within wide limits. Thus, the resultant direct reaction of a rotary converter may be adjusted to almost any desired value.

The ohmic drop in the armature of a rotary converter has a different expression than in either a direct-current or a synchronous machine, because the i^2r loss must be calculated for the actual shape of the superimposed currents. Rotary converters are sometimes provided with interpoles, in order to improve the commutation and to use a higher specific electric loading.[1]

[1] For further details in regard to the armature reaction in a rotary converter see Arnold, *Wechselstromtechnik*, Vol. 1 (1904), Chap. 28; Pichelmayer, *Dynamobau* (1908), p. 276; *Standard Handbook*, index under " Converter, synchronous, effective armature resistance "; Parshall & Hobart, *Electric Machine Design* (1906) p. 377; Lamme and Newbury, Interpoles in Synchronous Converters, *Trans. Amer. Inst. Electr. Engrs.*, Vol. 29 (1910), p. 1625.

CHAPTER X

ELECTROMAGNETIC ENERGY AND INDUCTANCE

56. The Energy Stored in an Electromagnetic Field. Experiment shows that no supply of energy is required to maintain a constant magnetic field. The power input into the exciting coil or coils is in this case exactly equal to that converted into the i^2r heat in the conductors, and this power is the same whether a magnetic field is produced or not. Another familiar example is that of a permanent magnet, which maintains a magnetic field without any supply of energy from the outside, and apparently without any decrease in its internal energy. Nevertheless, every magnetic field has a certain amount of energy stored within it, though in a form yet unknown. This is proved by the fact that an expenditure of energy is required to increase the field, and, on the other hand, when the flux is reduced, some energy is returned into the exciting electric circuit.

The conversion of electric into magnetic energy and *vice versa* is accomplished through the e.m.f. induced by the varying flux. Let a magnetic flux be excited by a coil CC (Fig. 45) supplied with current from a source of constant voltage E, and let there be a rheostat r in series with the coil. Let part of the resistance in the rheostat be suddenly cut out in order to increase the current in the coil. It will be found that the current rises to its final value not instantly; namely, when the current increases, the flux also increases, and in so doing it induces in the electric circuit an e.m.f., say e, which tends to oppose the current. This e, together with the iR drop, balances the voltage E, so that for a time the power Ei supplied by the source is larger than the power i^2R lost in the total resistance of the circuit. The difference, $Ei - i^2R$, is stored in the magnetic field created by the coil and by the other parts of the circuit. The energy $ei\,dt$ supplied by the electric circuit during the element of time dt is converted into the magnetic energy of the field, by the law of the conservation of

177

energy, while the amount $(E-e)idt=i^2R\,dt$ is converted into heat.

If now the same resistance is suddenly introduced into the circuit, the current gradually returns to its former value, and during this variable period the i^2R loss is larger than the power Ei supplied by the source. The applied voltage is in this case assisted by the voltage e induced by the decreasing field, the e.m.f. e supplying part of the i^2R loss.

To make the matter more concrete let the source of electric energy be a constant-voltage, direct-current generator, driven by

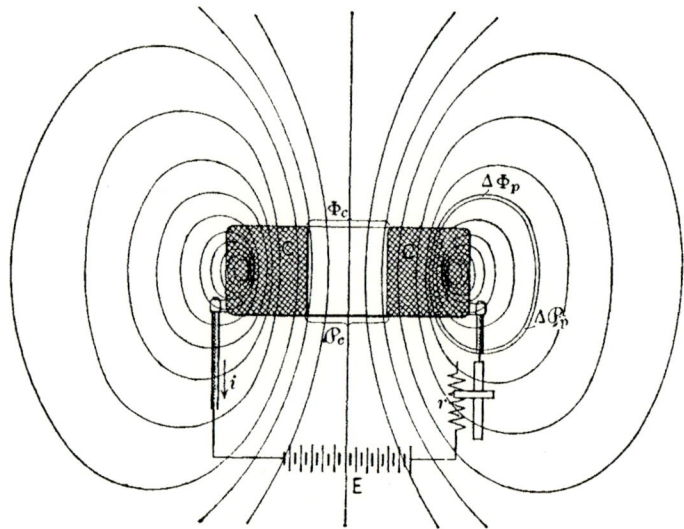

FIG. 45.—The magnetic field produced by a coil, showing complete and partial linkages.

a steam turbine. Let the load consist of coils of variable resistance R, and let the coils produce a considerable magnetic field. As long as the load current is constant, the rate of the steam consumption is determined by the i^2R loss in the circuit. When the current increases, the steam is consumed at each instant at a higher rate than it would be with a constant current of the same instantaneous value. The energy of steam is thus partly stored in the magnetic field of the coils. When the current is returned to its former value, the steam consumption during the transitional period is less than that which corresponds to the i^2R loss in the circuit, so that practically the same amount of steam is saved which was expended before in increasing the magnetic field.

These phenomena may be explained, or at least expressed in more familiar terms, by assuming the magnetic field to be due to some kind of motion in a medium possessed of inertia (Art. 3). When the field strength is increased it becomes necessary to accelerate the parts in motion, overcoming their inertia. When the field is reduced, the kinetic energy of motion is returned to the electric circuit. One can also conceive of the energy of the magnetic field to be static and in the form of some elastic stress. Under this hypothesis, when a current increases, the magnetic stress also increases at the expense of the electric energy. In either case, when the current is constant no energy is required to maintain the field, any more than to maintain a constant rotation in a fly-wheel or a constant stress in an elastic body.

It seems the more probable that the magnetic energy of a circuit is stored in some kinetic form, because the current which accompanies the flux is itself a kinetic phenomenon. On the other hand, it appears more likely that electrostatic energy is due to some elastic stresses and displacements in the medium, and thus it may be said to be potential energy. Electric oscillations and waves consist, then, in periodic transformations of electrostatic into electromagnetic energy, or potential into kinetic energy, and *vice versa*, similar to the mechanical oscillations and waves in an elastic body. In the familiar case of current or voltage resonance the total energy of the circuit at a certain instant is stored in the form of electrostatic energy in the condenser (permittor) connected into the circuit, or in the natural permittance of the circuit; the current and the magnetic energy at this instant are equal to zero. At another instant, when the current and the magnetic field are at their maximum, the energy stored is all in the form of magnetic energy, and the voltage across the condenser and the stress in the dielectric are equal to zero. An oscillating pendulum offers a close analogy to such a system. The resistance of the electric circuit, and the magnetic and dielectric hysteresis, are analogous to the friction and windage which accompanies the motion of the pendulum.

As it is of importance in mechanical machine design to know the inertia of the moving parts of a machine, so it is often necessary in the design and operation of electrical apparatus and circuits to know the amount of energy stored in the magnetic field, or the electro-magnetic inertia. This inertia modifies the current and voltage relations in the electric circuit in somewhat the same way in which the inertia of the reciprocating parts in an engine modifies the useful effort. In mechanical design a revolving part is characterized by its moment of inertia from which the stored energy can be calculated for any desired speed. So in electrical engineering a circuit or an apparatus is characterized by its electro-magnetic inertia or *inductance*, from which it is possible to calculate the magnetic energy stored in it at any desired value of the current. In this and in the two next chapters expressions are deduced for the inductance of some of the principal types of apparatus used in electrical engineering.

Prob. 1. A stationary electromagnet attracts and lifts its armature with a weight attached to it. Explain how the energy necessary for the lifting of the weight is supplied by the electric circuit.

Prob. 2. A direct-current generator driven by a water-wheel is subjected to very large and sudden fluctuations of the load, which the governor and the gate mechanism are unable to follow properly. A heavy fly-wheel on the generator shaft would improve the operating conditions. Would a reactance coil in series with the main circuit achieve the same result, if it could be made large enough?

Prob. 3. What experimental evidence could be offered to support the contention that the energy of an electric circuit is contained in the magnetic field linked with the current, and not in the current itself? The flow of current is usually compared to that of water in a pipe; is not all the kinetic energy stored in the moving water itself and not in the surrounding medium, and if so, is a current of electricity really like a current of water?

Prob. 4. Describe in detail current and voltage resonance[1] and free electrical oscillations, from the point of view of the periodic conversion of electromagnetic into electrostatic energy and *vice versa*, taking account of the dissipation of energy in the resistance of the circuit.

57. Electromagnetic Energy Expressed through the Linkages of Current and Flux.

In order to obtain a general expression for the energy stored in the magnetic field of an electric circuit, consider first a single loop of wire *aa* (Fig. 11) through which a steady electric current i is flowing. Let the cross-section of the wire be small as compared to the dimensions of the loop, so that the flux

---

[1] See the author's *Experimental Electrical Engineering*, Vol. 2, pp. 17 to 25; *The Electric Circuit*. Art. 67.

inside the wire may be disregarded. The electromagnetic energy possessed by the loop is equal to the electrical energy spent in establishing the current i in the loop against the induced e.m.f. Let i_t and Φ_t be the instantaneous values of the current in amperes and the flux in webers at a moment t during the period of building up the flux, and let e_t be the instantaneous applied voltage. Let the flux increase by $d\Phi_t$ during an infinitesimal interval dt; then the electrical energy (in joules) supplied from the source of power to the magnetic field is

$$dW = i_t e_t dt = i_t (d\Phi_t/dt) dt = i_t d\Phi_t,$$

where $e_t = d\Phi_t/dt$ is the instantaneous e.m.f. necessary to apply to the loop in order to counterbalance that induced by the changing flux. The total energy supplied from the electrical source during the period of building up the field to its final value Φ is

$$W = \int_0^i i_t d\Phi_t. \quad \ldots \ldots \quad (98)$$

In a medium of constant permeability the integration can be easily performed, because the flux is proportional to the current, or, according to eq. (2) in Art. 5, $\Phi_t = \mathcal{P} i_t$, where $\mathcal{P}$ is the permeance of the magnetic circuit, in henrys. Eliminating by means of this relation either i_t or Φ_t from eq. (98) we can obtain any one of the following three expressions for the electromagnetic energy stored in the loop:

$$\left. \begin{array}{l} W = \tfrac{1}{2} i \Phi; \\ W = \tfrac{1}{2} i^2 \mathcal{P}; \\ W = \tfrac{1}{2} \Phi^2/\mathcal{P}. \end{array} \right\} \quad \ldots \ldots \ldots \quad (99)$$

In the first form, eq. (99) expresses the fact that the magnetic energy stored in a loop is equal to one-half the product of the current by the flux; in the second form, it shows that the stored energy is proportional to the square of the current and to the permeance of the magnetic circuit. Both forms are of importance in practical applications.

Take now the more general case of a coil of n turns (Fig. 45); the flux which links with a part of the turns is now of a magnitude comparable with that of the flux which links with all the turns of the coil. We shall consider the *complete linkages* and the *partial linkages* separately. Consider first the energy due to the flux which links with all the turns of the coil. The e.m.f. induced in the coil by this flux, when the current changes, is $e_t = -n(d\Phi_t/dt)$, and the relation between the current and the flux is $\Phi_t = \mathcal{P}_c n i_t$, where $\mathcal{P}_c$ is the permeance of the path of the complete linkages. By repeating the reasoning given above in the case of a single loop we find that

$$W_c = \tfrac{1}{2}ni\Phi_c, \quad \ldots \ldots \ldots \quad (100)$$

or

$$W_c = \tfrac{1}{2}n^2 i^2 \mathcal{P}_c, \quad \ldots \ldots \ldots \quad (100a)$$

where the subcript c signifies that the quantities refer to the complete linkages only (Fig. 45). Two forms only are retained, being those that are of the most practical importance.

The energy of the partial linkages is calculated in a similar manner. Let $\Delta\Phi_p$ be a small annular flux (Fig. 45) which links with n_p turns of the coil, where n_p may be an integer or a fraction. For these turns the linkage with $\Delta\Phi_p$ is a complete linkage, while for the external $(n - n_p)$ turns it is no linkage at all and represents no energy, because no e.m.f. is induced by $\Delta\Phi_p$ in the turns external to it. Thus, the energy due to the flux $\Delta\Phi_p$, according to eqs. (100) and (100a), is equal to $\tfrac{1}{2}n_p i \Delta\Phi_p$, or to $\tfrac{1}{2}n_p{}^2 i^2 \Delta\Phi_p$. The total energy of the partial linkages is the sum of such expressions, over the whole flux Φ_p, or

$$W_p = \tfrac{1}{2}i\Sigma n_p \Delta\Phi_p, \quad \ldots \ldots \ldots \quad (101)$$

or

$$W_p = \tfrac{1}{2}i^2\Sigma n_p{}^2 \Delta\mathcal{P}_p. \quad \ldots \ldots \ldots \quad (101a)$$

The total energy of the coil is

$$W = \tfrac{1}{2}i[n\Phi_c + \Sigma n_p \Delta\Phi_p], \quad \ldots \ldots \quad (102)$$

or

$$W = \tfrac{1}{2}i^2[n^2\mathcal{P}_c + \Sigma n_p{}^2 \Delta\mathcal{P}_p], \quad \ldots \ldots \quad (102a)$$

where the first term on the right-hand side refers to the complete linkages and the second to the partial linkages of the flux and the current. In these expressions the current is in amperes, the fluxes in webers, the permeances in henrys, and the energy in joules (watt-seconds). If other units are used the corresponding numerical conversion factors must be introduced.

Some new light is thrown upon these relations by using the m.m.f. M instead of the ampere-turns ni. Namely, eqs. (102) and (102a) become:

$$W = \tfrac{1}{2}[M_c \Phi_c + \Sigma M_p \varDelta \Phi_p], \quad \ldots \quad (103)$$

or

$$W = \tfrac{1}{2}[M_c^2 \mathcal{P}_c + \Sigma M_p^2 \varDelta \mathcal{P}_p]. \quad \ldots \quad (103a)$$

These expressions are analogous to those for the energy stored in an electrostatic circuit, viz., $\tfrac{1}{2}EQ$, and $\tfrac{1}{2}E^2C$ (see *The Electric Circuit*). The m.m.f. M_c is analogous to the e.m.f. E; the magnetic flux Φ_c is analogous to the electrostatic flux Q, and the permeance $\mathcal{P}_c$ is analogous to the permittance C.

We can assume as a fundamental law of nature the fact that with a given steady current the magnetic field is distributed in such a way that the total electromagnetic energy of the system is a maximum. All known fields obey this law, and, in addition, it can be proved by the higher mathematics. Eq. (102a) shows that this law is fulfilled when the sum $n_c^2 \mathcal{P}_c + \Sigma n_p^2 \mathcal{P}_p$ is a maximum. When the partial linkages are comparatively small, the energy stored is a maximum when the permeance $\mathcal{P}_c$ of the paths of the total linkages is a maximum. This fact is made use of in the graphical method of mapping out a magnetic field, in Art. 41 above.

Prob. 5. The no-load saturation curve of an 8-pole electric generator is a straight line such that when the useful flux is 10 megalines per pole the excitation is 7200 amp.-turns per pole; the leakage factor is 1.2. Show that at this excitation there is enough energy stored in the field to supply a small incandescent lamp with power for a few minutes.

Prob. 6. Explain the function and the diagram of connections of a field-discharge switch.

Prob. 7. Prove that the magnetic energy stored in an apparatus containing iron is proportional to the area between the saturation curve and the axis of ordinates. The saturation curve is understood to give the total flux plotted against the exciting ampere-turns as abscissæ. Hint: See Art. 16.

Prob. 8. Deduce expression (102a) directly, by writing down an expression for the total instantaneous e.m.f. induced in a coil (Fig. 45).

Prob. 9. Explain the reason for which, in the formulæ deduced above, it is permissible to consider n to be a fractional number.

58. Inductance as the Coefficient of Stored Energy, or the Electrical Inertia of a Circuit. Eq. (102a) shows that in a magnetic circuit of constant permeability the stored energy is proportional to the square of the current which excites the field. The coefficient of proportionality, which depends only upon the form of the circuit and the position of the exciting m.m.f., is defined as the *inductance* of the electric circuit. The older name for inductance is the coefficient of self-induction. It is assumed here that the magnetic circuit is excited by only *one* electric circuit, so that there is no mutual inductance. Thus, by definition

$$W = \tfrac{1}{2}Li^2. \qquad \qquad (104)$$

where the inductance is

$$L = n^2 \mathscr{P}_c + \Sigma n_p{}^2 \!\varDelta \mathscr{P}_p, \qquad (105)$$

or, replacing the summation by an integration,

$$L = n^2 \mathscr{P}_c + \int_0^n n_p{}^2 d\mathscr{P}_p. \qquad (106)$$

Since the permeances in eq. (102a) are expressed in henrys, and the numbers of turns are numerics, the inductance L in the defining eqs. (105), or (106), is also in henrys. If the permeances are measured in millihenrys or in perms, the inductance L is measured in the same units. As a matter of fact, the henry was originally adopted as a unit of inductance, and only later on was applied to permeance.[1]

In some cases it is convenient to replace the actual coil (Fig. 45) by a fictitious coil of an equal inductance, and of the same number of turns, but without partial linkages. Let $\mathscr{P}_{eq}$ be the permeance of the complete linkages of this fictitious coil; then, by definition, eqs. (105) and (106) become

$$L = n^2 \mathscr{P}_{eq}. \qquad (106a)$$

This expression is used when the permeance of the paths is calculated from the results of experimental measurements of inductance, because in this case it is not possible to separate the partial linkages. Use is made of formula (106a) in chapter XII, in calculating the inductance of armature windings.

---

[1] The use of the henry as a unit of permeance was proposed by Professor Giorgi. See *Trans. Intern. Elec. Congress* at St. Louis (1904), Vol. 1, p. 136. The connection between inductance and permeance seems to have been first established by Oliver Heaviside; see his *Electromagnetic Theory* (1894), Vol. 1, p. 31.

The inductance L is related in a simple manner to the electromotive force induced in the exciting electric circuit when the current varies in it. Namely, the electric power supplied to the circuit or returned from the circuit to the source is equal to the rate of change of the stored energy, so that we have from eq. (104)

$$dW/dt = i(-e) = Li(di/dt),$$

or, canceling i,

$$e = -L(di/dt), \quad \ldots \ldots \quad (107)$$

The sign minus is used because e is understood to be the induced e.m.f. and not that applied at the terminals of the circuit. Therefore, when dW/dt is positive, that is, when the stored energy increases with the time, e is induced in the direction opposite to that of the flow of the current, and hence by convention is considered negative. Inductance is sometimes defined by eq. (107), and then eqs. (104) and (105) are deduced from it. The definition of L by the expression for the electromagnetic energy seems to be a more logical one for the purpose of this treatise, while the other definition in terms of the induced e.m.f. is proper from the point of view of the electric circuit.

Looking upon the stored magnetic energy as due to some kind of a motion in the medium, eq. (104) suggests the familiar expressions $\frac{1}{2}mv^2$ and $\frac{1}{2}K\omega^2$ for the kinetic energy of a mechanical system. Taking the current to be analogous to the velocity of motion, the inductance becomes analogous to mechanical mass and moment of inertia. The larger the electromagnetic inertia L the more energy is stored with the same current. Equation (107) also has its analogue in mechanics, namely in the familiar expressions mdv/dt and $Kd\omega/dt$ for the accelerating force and torque respectively. The e.m.f. e represents the reaction of the circuit upon the source of power when the latter tends to increase i the rate of flow of electricity. While these analogies should not be carried too far, they are helpful in forming a clearer picture of the electromagnetic phenomena.

The role of inductance, L, in the current and voltage relations of alternating-current circuits is treated in detail in the author's *Electric Circuit*. In this book inductance is considered from the point of view of the magnetic circuit, i.e., as expressed by eqs. (104) to (106). In the next two chapters the values of inductance are calculated for some important practical cases, from the forms and the dimensions of the magnetic circuits, using the fundamental equations (104), (105) and (106). The reader will see that the problem is reduced to the determination of various permeances and fluxes; hence, it presents the same difficulties with which he is already familiar from the study of Chapters V and VI.

Inductance of electric circuits in the presence of iron. When iron is present in the magnetic circuit, three cases may be distinguished:

(1) The reluctance of the iron parts is negligible as compared to that of the rest of the circuit;

(2) The reluctance of the iron parts is constant within the range of the flux densities used;

(3) The reluctance of the iron parts is considerable, and is variable.

In the first two cases, eqs. (104), (105) and (106) hold true, and the inductance can be calculated from the constant permeances of the magnetic circuit. In the third case, inductance, if used at all, must be separately defined, because eq. (102a) does not hold when the permeance of the circuit varies with the current. The equation of energy is in this case

$$W = n \int_0^t i_t d\Phi_{tc} + \Sigma \int_0^t n_p i_t d(\Delta\Phi_{tp}). \quad . \quad . \quad (108)$$

This equation is deduced by the same reasoning as eq. (98).

The following three definitions of inductance are used by different authors when the reluctance of a magnetic circuit is variable: (*a*) the expressions (104) and (108) are equated to each other, and L is calculated separately for each final value i of the current. Thus L is variable, and neither eq. (105) nor (107) hold true. (*b*) L is defined from eq. (107); in this case neither eq. (104) nor (105) are fulfilled. (*c*) L is defined at a given current by eq. (105) so that Li represents the sum of the linkages of the flux and the current. Therefore eq. (107) becomes $e = -d(Li)/dt$, and $dW = id(Li)$. With each of the three definitions L is variable, and therefore is not very useful in applications. The author's opinion is that when the permeance of the circuit is variable, L should not be introduced at all, but the original equation of energy (108) be used directly. Or else in approximate calculations, a constant value of L can be used, calculated for some average value of i or Φ.

Prob. 10. It is desired to make a standard of inductance of one millihenry by winding uniformly one layer of thin flat conductor upon a toroidal wooden ring of circular cross-section. How many turns are needed if the diameter of the cross-section of the ring is 10 cm. and the mean diameter of the ring itself is 50 cm.?   Ans. 400.

Prob. 11. An iron ring of circular cross-section is uniformly wound with n turns of wire, the total thickness of the winding being t; the mean diameter of the ring is D and the radius of its cross-section r. What is the inductance of the apparatus, assuming the permeability of the iron to be constant and equal to 1500 times that of the air? Hint: $d\mathcal{P}_p = \mu 2\pi(r+x)dx/\pi D$; $n_p = n(t-x)/t$.

Ans. $1.25(n^2/D) [1500r^2 + t(\tfrac{2}{3}r + \tfrac{1}{4}t)] \times 10^{-8}$ henry.

Prob. 12. A ring is made of non-magnetic material, and has a rectangular cross-section of dimensions b and h; the mean diameter of the ring is D. It is uniformly wound with n turns of wire, the total thickness of the winding being t. What is the inductance of the winding?

Ans. $0.4(n^2/D)[bh + \tfrac{2}{3}t(b+h+0.786t)]10^{-5}$ millihenry.

Prob. 13. The ring in the preceding problem has the following dimensions: $D = 50$ cm.; $h = b = 10$ cm.; it is to be wound with a conductor 3 mm. thick (insulated). How many turns are required in order to get an inductance of 0.43 of a henry? Hint: Solve by trials, assuming reasonable values for t; the number of turns per layer decreases as the thickness of the winding increases.   Ans. About 5600.

Prob. 14. It is desired to design a choke coil which will cause a reactive drop of 250 volts, at 10 amp. and 50 cycles. The gross cross-section of the core (Fig. 12) is 120 sq.cm., and the mean length of the path 130 cm.; the maximum flux density in the iron must be not over 7 kl. per sq. cm. What is the required number of turns and the length of the air-gap in the core?[1]   Ans. 149; 3.7 mm.

Prob. 15. An electrical circuit, which consists of a Leyden jar battery of 0.01 mf. capacity and of a coil having an inductance of 10 millihenrys, undergoes free electrical oscillations in such a way that the maximum instantaneous voltage across the condenser is 10,000 volts. What is the current through the inductance one-quarter of a cycle later, neglecting any loss of energy during the interval?

Ans. 10 amp.

Prob. 16. Suggest a practical experiment which would prove directly that the stored electromagnetic energy is proportional to the square of the current.

Prob. 17. Prove that the inductance of a coil of given external dimensions is proportional to the square of the number of turns, taking into account the complete and the partial linkages. Show that the ohmic resistance of the coil is also proportional to the square of the number of turns, provided that the space factor is constant.

NOTE 1. The theoretical calculation of the inductance of short

---

[1] For a complete design of a reactive coil see G. Kapp, *Transformers* (1908), p. 105.

straight coils and loops of wire in the air is rather complicated, because of the mathematical difficulties in expressing the permeances of the paths. Those interested in the subject will find ample information in Rosa and Cohen's *Formula and Tables for the Calculation of Mutual and Self-Inductance*, in the Bulletins of the Bureau of Standards. Vol. 5 (1908), No. 1. The article contains also quite a complete bibliography on the subject. See also Orlich, *Kapazität und Inductivität* (1909), p. 74 *et seq.*

Note 2. In the formulæ deduced in this and in the two following chapters, it is presupposed that the current is distributed uniformly over the cross-section of the conductors. Such is the case in conductors of moderate size and at ordinary commercial frequencies, unless perchance the material is itself a magnetic substance. With high frequencies, or with conductors of unusually large transverse dimensions, as also with conductors of a magnetic material, the current is not distributed uniformly over the cross-section of the conductors, the current density being higher near the periphery. The result is that, as the frequency increases, the inductance becomes lower and the ohmic resistance higher. This is known as the *skin effect*. For an explanation, for a mathematical treatment in a simple case, and for references see Heinke, *Handbuch der Elektrotechnik*, Vol. 1 (1904) part 2. pp. 120 to 129. Tables and formulæ will be found in the *Standard Handbook*, and in Foster's *Pocket-Book*. See also C. P. Steinmetz, *Alternating-Current Phenomena* (1908), pp. 206–208, and his *Transient Electric Phenomena* (1909), Section III, Chapter VII; Arnold, *Die Wechselstromtechnik*, Vol. 1 (1910), p. 564; A. B. Field, Eddy Currents in Large Slot-wound Conductors, *Trans. Amer. Inst. Electr. Engrs.*, Vol. 24 (1905), p. 761.

CHAPTER XI.

THE INDUCTANCE OF CABLES AND OF TRANS-
MISSION LINES.

59. The Inductance of a Single-phase Concentric Cable. Let Fig. 46 represent the cross-section of a concentric cable, which consists of an inner core A and of an external annular conductor D, with some insulation C between them. Let the radii of the conductors be a, b, and c, respectively. The insulation outside of D and the sheathing are not shown. Let a direct current of i amperes flow through the inner conductor away from the reader and return through the outer conductor. The magnetic field produced by this current links with the current, and for reasons of symmetry the lines of force are concentric circles. The field is confined within the cable, because outside the external conductor D the m.m.f. is $i-i=0$.

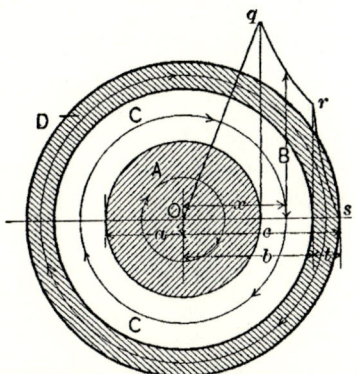

Fig. 46—The magnetic field within a concentric cable.

In the space between the two conductors the lines of force are linked with the whole current, and since there is but one turn, the m.m.f. is equal to i. The length of a line of force of a radius x cm. is $2\pi x$ so that the magnetic intensity is $H = i/2\pi x$, amp. turns per cm., the corresponding flux density $B = \mu i/2\pi x$ maxwells per sq.cm. Thus, the flux density decreases inversely as the distance from the center; it is represented by the ordinates of the part qr of a hyperbola.

In the space within the inner conductor A, a line of force of radius x is linked with a current $i_x = i(\pi x^2/\pi a^2) = i(x/a)^2$, provided

189

that the current is distributed uniformly over the cross-section of
the conductor. The length of the line of force is $2\pi x$, so that
$H = i_x/2\pi x = xi/(2\pi a^2)$, and $B = \mu x i/(2\pi a^2)$. Thus, in this part
of the field the flux density increases as the distance from the
center and is represented by the straight line Oq.

In the space inside the conductor D, a line of force of a
radius x is linked with the current $-i(x^2 - b^2)/(c^2 - b^2)$ of the
external conductor and with the current $+i$ of the internal conductor, or altogether with the current $i_x = i(c^2 - x^2)/(c^2 - b^2)$.
Consequently, here the flux density is represented by the
hyperbola rs, the equation of which is

$$B = \mu i_x/2\pi x = \mu i(c^2 x - x) [2\pi(c^2 - b^2)].$$

The curve $Oqrs$ gives a clear physical picture of the field distribution in the cable, and helps one to understand the linkages
which enter into the calculation of the inductance of the cable.

The inductance of the cable is calculated according to the
fundamental formula (106), the complete linkages being in the
space between the two conductors, and the partial linkages being
within the space occupied by the conductors themselves. Consider a piece of the cable one centimeter long. The permeance of a
tube of force of a radius x and of a thickness dx is $\mu dx/2\pi x$, so that
the permeance of the complete linkages is,

$$L_c' = \mathcal{P}_c' = \int_a^b \mu dx/2\pi x = (\mu/2\pi) \mathrm{Ln}(b/a) \text{ perm/cm.}, \quad (109)$$

where Ln is the symbol for natural logarithms. In this case the
permeance is equal to the inductance because the number of turns
$n = 1$. The sign " prime " indicates that the quantities L_c' and
$\mathcal{P}_c'$ refer to a unit length of the cable.

For the space inside the inner conductor $n_p = (x/a)^2$, this being
the fraction of the current with which the line of force of radius x
is linked. Hence, the part of the inductance of the cable due to
the field inside the conductor A is

$$L_A' = (\mu/2\pi) \int_0^a (x/a)^4 (dx/x) = \mu/8\pi = 0.05 \text{ perm/cm.} \quad (110)$$

This formula shows that the part of the inductance due to the
field within the inner conductor is independent of the radius of the

conductor, and is always equal to 0.05 perm. per cm., or 0.05 milli-henry per kilometer length of the cable.

The exact expression for the part of the inductance of the cable L_D' due to the linkage within the outer conductor is given in problem 10 below. The formula is rather complicated for prac-tical use, especially in view of the fact that this part of the induct-ance is comparatively small, because the flux density on the part rs of the curve is small. It is more convenient, therefore, to make simplifying assumptions, when the thickness t of the outer con-ductor is small as compared to b. Namely, the length of all the paths within the outer conductor may be assumed to be equal to $2\pi b$, so that the permeance of an infinitesimal path of a radius x and thickness dx nearly equals $\mu dx/2\pi b$. Furthermore, the volume of the current in the outer conductor, between the radii b and x may be assumed to be proportional to the distance $x-b$, and hence equal to $i(x-b)/(c-b)$. A line of force of a radius x is linked therefore with the whole current i in the inner conductor and with the above-stated part of the current in the outer conductor, and, since the currents flow in opposite directions, this line of force is linked altogether with the current $i(c-x)/(c-b)$. Hence, it is linked with $n_p = (c-x)/(c-b)$ turns. Thus, the inductance of the cable, due to the outer partial linkages, is, in the first approxi-mation,

$$L_D' = \mu/(2\pi b t^2) \int_b^c (c-x)^2 dx = \tfrac{1}{15}t/b \text{ perm/cm.} \qquad . \quad (111)$$

If a closer approximation is desired, it is convenient to expand eq. (114) in ascending powers of t/b, as is explained in problem 11. The result is

$$L_D' = \tfrac{1}{15}t/b[1 - \tfrac{1}{10}(t/b)^2 + \tfrac{3}{40}(t/b)^3 \ldots] \text{ perm/cm.} \quad . \quad (112)$$

It will thus be seen that eq. (111) is an accurate approximation, because eq. (112) contains in the parentheses no term with the first power of the ratio t/b.

Thus, the total inductance of a concentric cable, l kilometers long, is

$$L = [0.46 \log_{10} (b/a) + 0.05 + L_D']l \text{ millihenrys,} \quad . \quad (113)$$

where L_D' is given by eqs. (111), (112), or (114), according to the accuracy desired. Expressions (110) and (114) are correct only at low frequencies, such as are used for power transmission. With very high frequencies, the skin effect becomes noticeable, that is, the current in the inner conductor is forced outward and that in the outer conductor inward. In the limit, when the frequency is infinite, the currents are concentrated on the opposing surfaces of the conductors, and the partial linkages are equal to zero. Thus, each of the expressions in question must be multiplied by a variable coefficient k which, for a given cable, is a function of the frequency. At ordinary frequencies $k = 1$, and gradually approaches zero as the frequency increases to infinity.[1]

Prob. 1. A concentric cable is to be designed for 750 amperes, the current density to be about 2.2 amp. per gross sq.mm., and the thickness of the insulation between the conductors to be 6 mm. What are the dimensions of the conductors assuming them to be solid, that is, not stranded? The fact that they are in reality stranded is taken care of in the permissible current density.

Ans. $a = 10.5$; $b = 16.5$; $c = 19.6$ mm.

Prob. 2. Plot the curve $0qrs$ of distribution of the flux density in the cable given in the preceding problem.

Ans. At $x = a$, $B = 143$: at $x = b$, $B = 91$ maxwells per sq.cm.

Prob. 3. What is the total flux in megalines per kilometer of the cable specified in the two preceding problems?   Ans. 15.1.

Prob. 4. Show how to plot the curve of the distribution of flux density in a three-phase concentric cable, at some given instantaneous values of the three currents.

Prob. 5. What is the inductance of a 25-km. cable in which the diameter of the inner conductor is 12 mm., the thickness of insulation is 3 mm., and the dimension c is such that the current density in the outer conductor is 10 per cent higher than in the inner one?

Ans. $25[0.0810 + 0.050 + 0.0122] = 3.58$ millihenry.

Prob. 6. A cable consists of three concentric cylinders of negligible thickness; the radii of the cylinders are r_1, r_2, and r_3, beginning with the inner one. What is the inductance in millihenrys per kilometer, when a current flows through the inner cylinder and returns equally divided through the two others?

Ans. $0.46 [\log (r_2/r_1) + 0.25 \log (r_3/r_2).]$

Prob. 7. In the cable given in the preceding problem the total current i flows through the middle cylinder, the part mi returns through the inner cylinder, and the rest, ni, returns through the outer one. What is the total inductance per kilometer of length?

Ans. $0.46 [m^2 \log (r_2/r_1) + n^2 \log (r_3/r_2)].$

[1] For the field distribution in and the inductance of non-concentric cables see Alex. Russell, *Alternating Currents*, Vol. 1 (1904), Chap. XV; for the reactance of armored cables see J. B. Whitehead, " The Resistance and React- ance of Armored Cables, *Trans. Amer. Inst. Electr. Engrs.*, Vol. 28 (1909), p. 737.

Prob. 8. At what ratio of b to a in Fig. 46 is the magnetic energy stored within the inner conductor equal to that stored between the two conductors?   Ans. 1.28.

Prob. 9. It is required to replace the solid inner conductor A in Fig. 46 by an infinitely thin shell of such a radius a' that the total inductance of the cable shall remain the same. What is the radius of the shell? Hint: $(\mu/2\pi)\mathrm{Ln}(a/a') = \mu/8\pi$.

Ans. $a'/a = \varepsilon^{-0.25} = 0.779.$

Prob. 10. Prove that the part of the inductance due to the linkages within the outer conductor in Fig. 46 is expressed by

$$L_D' = \frac{\mu}{2\pi(c^2-b^2)^2}[c^4\mathrm{Ln}(c/b) - \tfrac{1}{4}(3c^4 + b^4 - 4c^2b^2)]. \quad . \quad (114)$$

Hint: $d\mathcal{P}_p = \mu dx/2\pi x;\; n_p = 1 - \pi(x^2-b^2)/\pi(c^2-b^2).$

Prob. 11. Deduce eq. (112) from formula (114), assuming the ratio t/b to be small as compared to unity. Hint: Put $c = b(1+y)$ where $y = t/b$ is a small fraction. Expand $\mathrm{Ln}(1-y)$ into an infinite series, and omit in the numerator of eq. (114) all the terms above y^5; expand (c^2-b^2) in the denominator in ascending powers of y, and divide the numerator by this polynomial.

60. The Magnetic Field Created by a Loop of Two Parallel Wires.

Let Fig. 47 represent the cross-section of a single-phase or direct-current transmission line, the wires being denoted by A and B. With the directions of the currents in the wires shown by the dot and the cross, the magnetic field has the directions shown by the arrow-heads, one-half of the flux linking with each wire. Before calculating the inductance of the loop it is instructive to get a clear picture, quantitative as well as qualitative, of the field itself.

The field distribution is symmetrical with respect to the line AB and the axis OO'. The whole flux passes in the space between the wires, so as to be linked with the m.m.f. which produces it, and then extends to infinity on all sides. The flux density is at its maximum near the wires and gradually decreases toward OO' and toward $\pm \infty$, as is shown by the curve $pqsts'q'p'$. The ordinates of this curve represent the flux densities at the various points of the line passing through the centers of the wires. The reason for which the flux density is larger near the wires is that the path there is shorter, although the m.m.f. acting along all the paths is the same. This m.m.f. is numerically equal to the current i in the wires, the number of turns being equal to one.

It is proved below that the magnetic paths outside the conductors themselves are eccentric circles, with their centers on the line AB extended. The equipotential surfaces are circular cylinders, which are shown in Fig. 47 as circles passing through the centers A and B of the conductors. Within the conductors themselves there are no equipotential surfaces.

For purposes of analysis it is convenient to regard the field in Fig. 47 as the result of the superposition of two simpler fields, similar to those of the concentric cable of the preceding article. Con-

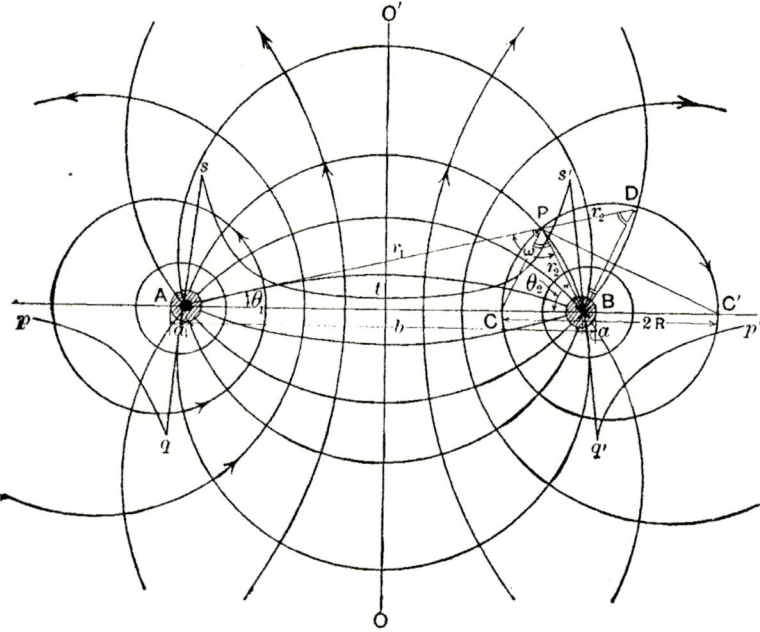

Fig. 47.—The magnetic field produced by a single-phase transmission line.

sider the conductor A, together with a concentric cylinder of an infinitely large radius, as one conducting system. Let the current flow through A toward the reader, and return through the infinite cylinder. Let the conductor B with a similar concentric cylinder form another independent system. The currents in the conductors A and B are to be the same as the actual currents flowing through them, but each infinite cylinder is to serve as a return for the corresponding conductor, as if there were no electrical connection between A and B. The currents in the two cylinders are flowing in opposite directions and the cylinders themselves coincide at infinity, because the distance AB between their axes is infinitely small as compared to their radii. Hence, the two currents in the cylinders cancel each other, and the combination of the two component systems is magnetically identical with the given loop.

In a medium of constant permeability, the resultant magnetic intensity H, produced at a point by the combined action of several independent m.m.fs., is equal to the geometric sum of the intensities produced at the same point by the separate m.m.fs.[1] This being true of the intensities, the component flux densities at any point are also combined according to the parallelogram law because they are proportional to the intensities. Hence, the resultant flux can be regarded as the result of the superposition of the fluxes created by the component systems.

The field produced by the system A consists of concentric circles, the flux density outside the conductor A being inversely proportional to the distance from the center of A (curve qr in Fig. 46). The field created by the system B consists of similar circles around B, and the field shown in Fig. 47 is a superposition of these two fields. Thus the resultant field intensity H at a point P is a geometric sum of

$$H_1 = i/2\pi r_1, \quad \ldots \ldots \quad (115)$$

and

$$H_2 = i/2\pi r_2, \quad \ldots \ldots \quad (116)$$

H_1 and H_2 being perpendicular to the corresponding radii vectors r_1 and r_2 from the centers of the conductors to the point P. The directions of H_1 and H_2 are determined by the right-hand screw rule. Since H_1 and H_2 are known in magnitude and direction at each point of the field, the resultant intensity H may also be determined.

To deduce the equation of the lines of force in the resultant field, we shall express analytically the condition that the total flux which crosses the surface CP is equal to zero, provided that C and P lie on the same line of force. This total flux may be considered

[1] This *principle of superposition* can be considered (*a*) as an experimental fact; (*b*) as an immediate consequence of the fact that in a medium of constant permeability the effects are proportional to the causes; (*c*) as a consequence of Laplace's law $dH = \text{Const.} \times i \, ds \, \sin \theta / 10r^2$, according to which the total field intensity is regarded as the sum of those produced by the infinitesimal elements of the current, or currents.

as the resultant of the fluxes due to the systems A and B. According to eq. (115), the flux density due to A, at a distance x from A, is $B_1 = \mu i/2\pi x$, so that the flux due to A which crosses CP is

$$\Phi_1' = \int_{AC}^{r_1} B_1 dx = (\mu i/2\pi)\, \mathrm{Ln}(r_1/AC)\,\mathrm{maxwells/cm.} \qquad . \quad (117)$$

This flux is directed *to the left*, looking from C to P. By analogy, the flux due to the system B is

$$\Phi_2' = (\mu i/2\pi)\, \mathrm{Ln}(r_2/BC)\,\mathrm{maxwells/cm.,} \qquad . \quad . \quad . \quad (118)$$

and is directed *to the right*, looking from C to P. The condition that no flux crosses CP is, that Φ_1' is equal to Φ_2', or

$$\mathrm{Ln}(r_1/AC) = \mathrm{Ln}(r_2/BC),$$

or

$$r_1/r_2 = AC/BC = \mathrm{Const.} \qquad . \quad . \quad . \quad . \quad (119)$$

This is the equation of a line of force in "di-polar" co-ordinates; the curve is such that the ratio of r_1 to r_2 remains constant. However, this constant is different for each line of force, because each line has its own point C'.

Eq. (119) may be proved to represent a circle, by selecting an origin, say at A, and substituting for r_1 and r_2 their values in terms of the rectangular co-ordinates x and y. The following proof by elementary geometry leads to the same result. Produce AP and lay off $PD = PB = r_2$. According to eq. (119), BD is parallel to CP, and consequently PC bisects the angle $APB = \omega$. Let the point C' lie on the same line of force with C; then no flux passes through PC'', and by analogy with eq. (119) we have

$$r_1/r_2 = AC''/BC'' = \mathrm{Const.} \qquad . \quad . \quad . \quad . \quad (120)$$

By plotting $PD' = r_2$ (not shown in figure) along PA, in the opposite direction from PD, and connecting D' to B, one can show as before that PC'' bisects the angle $BPD = 180° - \omega$. But the bisectors of two supplementary angles are perpendicular to each other; consequently, CPC'' is a right angle, and the point P lies on a semicircle erected on the diameter CC'. This semicircle is the line of force itself, because all the points, such as P, which are determined by C and C' must lie on it. Another semicircle below the line AB closes the line of force.

From eqs. (119) and (120) the following expressions are obtained for the radius R of the line of force:

$$R = \frac{BC}{1 - (BC/AC)}, \quad \cdots \cdots \quad (121)$$

or

$$R = \frac{BC'}{1 + BC'/AC'}, \quad \cdots \cdots \quad (122)$$

so that the line of force can be easily drawn for a given C or C'.

To prove that the equipotential lines are also circles we proceed as follows. The line AB is evidently an equipotential line, because it is perpendicular to all the lines of force. The difference of magnetic potential or the m.m.f. between AB and P, contributed by the system A, is equal to $i(\theta_1/2\pi)$ ampere-turns, where the angle BAP is denoted by θ_1. This is because the m.m.f. due to the system A, taken around a complete circle concentric with A, is equal to i, and is distributed uniformly along the circle, for reasons of symmetry. Or else it follows directly from eq. (115). By analogy, the difference of magnetic potential between AB and P, due to the system B, is $i(\theta_2/2\pi)$. Thus the total difference of potential between AB and P is

$$M_{cp} = (i/2\pi)(\theta_1 + \theta_2) = (i/2\pi)(\pi - \omega). \quad \cdots \quad (123)$$

This shows that the m.m.f. between any two points in the field is proportional to the difference in the angles ω at which the line AB is visible from these points. For any two points on the same equipotential line M_{cp} is the same, so that the equation of such a line is

$$\omega = \text{const.} \quad \cdots \cdots \cdots \quad (124)$$

This represents the arc of a circle passing through A and B, and corresponding to the inscribed angle ω.

Prob. 12. A single-phase transmission line consists of two conductors 1 cm. in diameter, and spaced 100 cm. between the centers. Draw the curve of flux density distribution (*pqst* in Fig. 47) for an instantaneous value of the current equal to 100 amp.

Ans. $x = 50.0 \quad 25.0 \quad 0.5 \quad -0.5 \quad -50.0 \quad -\infty$ cm.;
$B = 0.80 \quad 1.07 \quad 40.20 \quad -39.80 \quad -0.26 \quad 0$ maxw./sq. cm.

Prob. 13. For the transmission line in problem 12 draw the lines

of force which divide the total flux between the wires into ten equal
parts (not counting the flux within the wires).

  Ans. The circles nearest to OO' cross AB at a distance of 48.4
cm. from each other.

  Prob. 14. Referring to the two preceding problems, draw ten
equipotential circles which divide the whole m.m.f. of 100 ampere-turns
into ten equal parts.

  Ans. The arcs nearest to AB intersect OO' at a distance of
32.5 cm. from each other.

  Prob. 15. A telephone line runs parallel to a single-phase power
transmission line. The position of one of the telephone wires is fixed;
show how to determine the position of the other wire so as to have a
minimum of inductive disturbance in the telephone circuit. Hint:
The center lines of the two telephone wires must intersect the same line
of force due to the power line.

  Prob. 16. A telephone line runs parallel to a 25-cycle, single-phase
transmission line. The distances from one of the telephone wires to
the power wires are 3.5 m. and 2.7 m.; the distances from the other
telephone wire to the power wires are 3.6 and 2.5 m. (in the same order).
What voltage is induced in the telephone line per kilometer of its length,
when the current in the power line is 100 effective amperes?

  NOTE: In practice, this voltage is neutralized by transposing
either line after a certain number of spans. Ans. 0.33 volt.

61. The Inductance of a Single-phase Line.

The inductance of
a single-phase line (Fig. 47) can be calculated according to the fundamental formulæ (105) or (106), provided that the permeances
of the elementary paths be expressed analytically. However, in
this case it is much simpler to use the principle of superposition
employed in the preceding article, and to consider the actual flux
as the resultant of two fluxes each surrounding concentrically one
of the wires and extending to infinity. The fluxes produced by
the two component systems are equal and symmetrical with
respect to the wires. It is therefore sufficient to calculate the
linkages of the loop AB with the flux produced by one of the systems, say that corresponding to A, and to multiply the result by
two.

  The flux produced by A, and having A as a center, links with
the current in the loop AB. These linkages may be divided into
the following:

  (a) Linkages within the wire A; that is, from $x = 0$ to $x = a$;

  (b) Linkages between the wires A and B, that is, from

$$x = a \text{ to } x = b - a;$$

(c) Linkages outside the loop, that is, from $x = b + a$ to $x =$ infinity,

(d) Linkages within the wire B, that is, from

$$x = b - a \text{ to } x = b + a.$$

The linkages (a), (b) and (c) are the same as in a concentric cable (Fig. 46), because the shape of the lines of force and the number of turns with which they are linked are the same. The partial linkages (d) are somewhat difficult to express analytically. When the distance b between the wires is large as compared to their diameters, the whole current in B may be assumed to be concentrated along the axis of the wire B, instead of being spread over the cross-section. With this assumption, the partial linkages (d) are done away with, the linkages (b) are extended to $x = b$, and the linkages (c) begin from $x = b$. The expressions for the linkages (a) and (b) are given by eqs. (110) and (109) respectively. The linkages (c) are equal to zero, because in this region the lines of force produced by A are linked with both A and B, and therefore with $i - i = 0$ ampere-turns. Thus, the inductance in question is

$$L' = 0.46 \log_{10}(b/a) + 0.05. \quad \ldots \quad (125)$$

This gives the inductance of a single-phase line in perms per centimeter length, or in millihenrys per kilometer length of the wire.[1] To obtain the inductance per unit length *of the line* this expression must be multiplied by two, because the linkages due to the flux produced by the system B are not taken into account in eq. (125). However, for the purposes of the next two articles it is more convenient to use expression (125), and to consider separately the inductance of each wire, remembering that the two wires of a loop are in series, and that therefore their inductances are added.[2]

Prob. 17. Check by means of formula (125) some of the values for the inductance and reactance of transmission lines tabulated in the various pocketbooks and handbooks.

[1] It is of interest to note that the exact integration over the partial linkages (d) leads to the same Eq. (125), so that this formula is correct even when the wires are close to each other. See A. Russell, *Alternating Currents*, Vol. 1 (1904), pp. 59–60.

[2] The inductance of two or more parallel cylinders of any cross-section can be expressed through the so-called " geometric mean distance," introduced by Maxwell. For details see Orlich, *Kapazität und Induktivität* (1909), pp. 63–74.

Prob. 18. Show by means of tables or curves that the inductance of a transmission line varies much more slowly than (a) the spacing with a given size of wire, (b) the size of wire with a given spacing.

Prob. 19. When the diameters of the two wires A and B are different, prove that the inductance of the complete loop is the same as if the diameter of each wire was equal to the geometric mean of the actual diameters.

Prob. 20. Show that the inductance of a single-phase line with a ground return can be calculated from eq. (125) by putting $b = 2h$ where h is the elevation of the wire above the ground. Hint: In Fig. 47 the plane OO' may be considered to be the surface of the earth, assumed to be a perfect conductor. If the earth be removed, an "image" conductor B must be added in order to provide a return path for the current, such that the field surrounding A would remain the same.

Prob. 21. Two single-phase lines are placed on two cross-arms of the same tower, one directly above the other, at a vertical distance of c cm apart. What is the total inductance of the combination, when the two lines are connected in parallel and each line carries one-half of the total current?

Solution: Consider the four wires as forming four fictitious systems, with cylinders at infinity as returns. Let b be the spacing in each loop, and let b be larger than c. Denote the wires in one loop by 1 and 2, in the other by 1' and 2', and let d be the diagonal distance between 1 and 2'. Assume all the wires except 1 to be of an infinitesimal cross-section. Then, the linkages of the flux produced by the system 1 with the currents in the four wires are

$$i^2 L_1' = 0.05(\tfrac{1}{2}i)^2 + 0.2(\tfrac{1}{2}i)^2 \operatorname{Ln}(c/a) + 0.2i\,(\tfrac{1}{2}i)\operatorname{Ln}(b/c)$$
$$+ 0.2(\tfrac{1}{2}i)^2 \operatorname{Ln}(d/b) \text{ millijoules/km.}$$

Thus, allowing the same amount for the linkages due to the current in the wire 1', we get that the inductance of the split line, each way, is

$$L' = 2L_1' = \tfrac{1}{2}[0.46 \log_{10}(bd/ca) + 0.05] \text{ millihenrys/km..}$$

instead of the expression (125) for the single line. The same result is obtained when b is smaller than c. Hence, by splitting a line in two the inductance is considerably reduced, because partial linkages are substituted for some of the complete linkages. If d were equal to c the reduction would be 50 per cent; but since d is always larger than c the gain is less than 50 per cent. However, when the two lines are very far apart the saving is very nearly 50 per cent.

Prob. 22. A certain single-phase transmission line has been designed to consist of No. 000 B. & S. conductor with a spacing of 180 cm. It is desired to reduce the reactive drop by about 20 per cent, without increasing the weight of copper, by using two lines in parallel, with the same spacing. What is the size of the conductor and the distance between the loops?   Ans. No. 1 B. & S.; about 8 cm.

Prob. 23. Solve problem 21 when the load is divided unequally between the two loops, the currents being mi and ni respectively, where $m+n=1$.

Ans. $L' = 0.46[(m^2+n^2) \log (c/a) + \log (b/c) + 2mn \log (d/b)] + 0.05(m^2+n^2)$, when $b>c$, and $L' = 0.46[(m^2+n^2) \log (b/a) + 2mn \log (d/c)] + 0.05(m^2+n^2)$, when $b<c$.

Prob. 24. A single-phase line consists of three conductors, the total current flowing through conductor 1, and returning through conductors 2 and 3 in parallel. If the current in one return conductor is mi, and in the other ni, where $m+n=1$, prove that the inductance of the line per kilometer of its length is, in millihenrys,

$$L' = 0.46 [\log (b_{13}/a_1) + m^2 \log (b_{23}/a_2) + n^2 \log (b_{23}/a_3) + m \log (b_{12}/b_{13}) + m \log (b_{12}/b_{23}) + n \log (b_{13}/b_{23})] + 0.05(1+m^2+n^2), \text{ when } b_{12}>b_{13}>b_{23}.$$

In the particular case when $b_{12}=b_{23}=b_{13}$, $a_1=a_2=a_3$, and $m=n=\frac{1}{2}$, the inductance is reduced by 25 per cent as compared to that of a single loop.[1]

62. The Inductance of a Three-phase Line with Symmetrical and Semi-symmetrical Spacing.

The magnetic field which surrounds a single-phase line varies in its intensity from instant to instant, as the current changes, but the direction of the magnetic intensity and of the flux density at each point remains the same. In other words, the flux is a pulsating one. The field created by three-phase currents in a transmission line varies at each point in both its magnitude and direction. At the end of each cycle, the field assumes its original magnitude and direction. If the spacing of the wires is symmetrical, the field at the end of each third of a cycle has the same magnitude and position with respect the next wire. The field may therefore be said to be revolving in space.

This revolving flux, like that in an induction motor, induces e.m.fs. in the three phases. The problem is to determine these counter-e.m.fs. in the transmission line, knowing the size of the wires, the spacing, and the load. In transmission line calculations, especially in determining the voltage drop and regulation, it is convenient to consider each wire separately, and to determine the voltage drop in phase and in quadrature with the current. Thus, having expressed the e.m.fs. induced by the revolving flux in terms of the constants of the line, each wire is then considered as if it were brought outside the inductive action of the two other wires.

We shall consider first the case of an equidistant spacing of the three wires, because in most practical calculations of voltage drop

---

[1] The splitting of conductors discussed in problems 21 to 24 has been proposed for extremely long transmission lines, in order to reduce their inductance and at the same time increase their electrostatic permittance (capacity).

an unsymmetrical spacing is replaced by an equivalent equidistant spacing. The exact solution for an unsymmetrical spacing is given in the next article. Let the instantaneous values of the three currents in the wires A, B and C of a three-phase transmission line be i_1, i_2 and i_3. The sum of the three currents at each instant is zero, or

$$i_1 + i_2 + i_3 = 0. \qquad \ldots \ldots \quad (126)$$

Let Φ_{eq} be the equivalent flux which links at any instant with the wire A. The instantaneous e.m.f. induced in this wire is

$$e_1 = -d\Phi_{eq}/dt. \qquad \ldots \ldots \quad (127)$$

The equivalent flux consists of the actual flux outside the wire plus the sum of the fluxes inside the wire, each infinitesimal tube of flues being reduced in the proper ratio, according to the fraction of the cross-section of the wire with which it is linked. Or, what is the same thing, each wire is replaced by an equivalent hollow cylinder of infinitesimal thickness, without partial linkages, as in problem 9 in Art. 59 (consult also the definition of equivalent permeance given in Art. 58).

In order to determine Φ_{eq} we replace the three-phase system by two superimposed single-phase systems. The current i_1 in the wire A may be thought of as the sum of the currents $-i_2$ and $-i_3$, each flowing in a separate fictitious wire, and both of these wires coinciding with A. The currents $+i_2$ in B and $-i_2$ in A form one single-phase loop, while the currents $+i_3$ in C and $-i_3$ in A form the other loop. The flux Φ_{eq} which surrounds A is the sum of the fluxes produced by these two loops. The flux per unit length of the line, due to the first loop, is equal to $-\mathcal{P}_{,q}' i_2$, since the number of turns is equal to one. For the same reason $\mathcal{P}_{,q}' = L'$ where L' is determined by eq. (125). Hence, the flux per unit length of the line, due to the first loop, is $-L' i_2$. Similarly, the flux due to the second loop and linked with A is equal to $-L' i_3$, the same value of L' being used because the spacing and the sizes of all of the wires are the same. Thus,

$$\Phi_{eq} = -L' i_2 - L' i_3 = L' i_1, \qquad \ldots \ldots \quad (128)$$

the last result being obtained by substituting the value of $i_2 + i_3$ from eq. (126). Thus, eq. (127) becomes simply

$$e_1 = -L' di_1/dt, \qquad \ldots \ldots \quad (129)$$

that is, the induced e.m.f. is the same as in a single-phase line carrying the current i_1. Thus, *the inductance of a three-phase line with symmetrical spacing, per wire, is the same as the inductance of a single-phase line, per wire, with the same size of wire and the same spacing.* The total e.m.f. induced in each wire is in quadrature with the current in the wire.

In reaching this conclusion the following facts were made use of: (*a*) The current in each wire at any instant is equal to the sum of the currents in the two other wires; (*b*) the fluxes due to separate m.m.fs. can be superimposed in a medium of constant permeability; (*c*) The inductance of the loop A-B is equal to that of A-C because of the same spacing. No other suppositions in regard to the character of the load or the voltages between the wires were made. Therefore, the conclusion arrived at holds true:

(*a*) For balanced as well as unbalanced loads;

(*b*) For balanced or unbalanced line voltages;

(*c*) For a three-wire two-phase system, three-wire single-phase system, monocyclic system, etc.

(*d*) For sinusoidal voltages as well as for those departing from this form.

Semi-symmetrical Spacing. When two out of the three distances between the wires in a three-phase line are equal to each other, the arrangement is called semi-symmetrical. Two common cases of this kind are: (*a*) When the wires are placed at the vertices of an isosceles triangle; (*b*) when they are placed at equal distances in the same plane, for instance on the same cross-arm, or are fastened to suspension insulators, one above the other. In such cases the inductive drop in the symmetrically situated wire is the same as if the wire belonged to a single-phase loop, carrying the same current, and with a spacing equal to the distance of this wire to either of the other two wires. Let, for instance, the distance A-B be equal to B-C, and let the distance A-C be different from the two. The proof given above can be repeated for the wire B, and the same conclusion will be reached because the spacing A-C is not used in the deduction. But, of course, the proof does not hold true for either wire A or C.

When the three wires are in the same plane, the inductance of each of the outside wires is larger than that of the middle wire. This can be shown as follows: Let the three wires be in a horizontal plane, and let them be denoted from left to right by A, B, C. Let the distance between A and B be equal to b and the distance between A and C be equal to $2b$. If the wire B were moved to coincide with C, the inductance of A would be the same as if it belonged to a single-phase loop with a spacing $2b$. If C were moved to coincide with B, the inductance of A would be that of a wire in a single-phase loop with a spacing b. Thus, the inductance of A corresponds in reality to a spacing intermediate between b and $2b$. The inductance of the middle wire B is the same as that of a wire in a single-phase loop with the spacing b, as is proved above. Thus, the inductance of either A or C is larger than that of B.

An inspection of a table of the inductances or reactances of transmission lines will show that the inductance increases much more slowly than the spacing. For instance, according to the *Standard Handbook*, the reactance per mile of No. 0000 wire, at 25 cycles, is 0.303 ohm with a spacing of 72 inch, and is 0.340 ohm with a spacing of 150 inch. Therefore, in practical calculations, when the spacing is semi-symmetrical, the values of inductance are taken the same for all the three wires, for an average spacing between the three, or, in order to be on the safe side, for the maximum spacing. In the most unfavorable case, even if an error of say 10 per cent be made in the estimated value of the inductance, and if the inductive drop is say 20 per cent of the load voltage, the error in the calculated value of voltage drop is only 2 per cent of the load voltage, and that at zero power factor. At power factors nearer unity, when the vector of the inductive drop is added at an angle to the line voltage, the error is much smaller.

Prob. 25. Show that the instantaneous electromagnetic energy stored per kilometer of a three-phase line with symmetrical spacing is equal to $\frac{1}{2}L'(i_1^2 + i_2^2 + i_3^2)$ millijoules per kilometer, where L' has the value given by eq. (125). If this is true, then each wire may be considered as if it were subjected to no inductive action from the other wires and had an inductance L' expressed by eq. (125). This is another way of proving eq. (129), and the statement printed in italics above. Solution: Consider each wire, with a concentric cylinder at infinity, as a component system. Determine the linkages of the field created by the system A with the currents in A, B, and C, as in Art. 61. The result is equal to $\frac{1}{2}L'i_1^2$. Similar expressions are then written by analogy for the fluxes due to the systems B and C.

Prob. 26. Show graphically that, when the distances A–B and A–C are equal, the equivalent flux linking with A is independent of the spacing B–C, and is the same as if B and C coincided. That is, prove that the inductance of A is the same as if it belonged to a single-phase loop. Solution: Let I_1, I_2, and I_3 (Fig. 48), represent the vectors of the three currents at an unbalanced load. The current I_1 in A is replaced by $-I_2$ and $-I_3$, and the system is split into two single-phase loops, A-B and A-C. The fluxes due to these systems and linked with A are denoted by Φ_{12} and Φ_{13}. They are in phase with the corresponding currents, and are proportional to the magnitudes of these currents, *because of the equal spacing*. Hence, the triangles of the currents and of the fluxes are similar, and the resultant flux Φ_1 linking with A is in phase with I_1. If $\Phi_{12} = \frac{1}{2}\mathcal{P}I_2$, and $_1\Phi_3 = \frac{1}{2}\mathcal{P}I_3$, where $\mathcal{P}$ is the equivalent permeance of each single-phase loop, then the result shows that $\Phi_1 = \frac{1}{2}\mathcal{P}I_1$. If the wires B and C coincided the equivalent permeance $\mathcal{P}$ would be the same, and hence the proposition is proved. The voltage drop, E_1, due to the flux Φ_1 is shown by a vector leading I_1 by 90 degrees.

63. The Equivalent Reactance and Resistance of a Three-phase Line with an Unequal Spacing of the Wires. In the case of an unequal spacing of the wires eqs. (126) and (127) still hold true, because they do not depend upon the spacing; but eq. (128) becomes

$$\Phi_{eq} = -L'_{12}i_2 - L'_{13}i_3, \quad \dots \quad (130)$$

where L'_{12} is the value of the inductance per unit length, calculated by eq. (125) for the spacing between A and B, and L'_{13} is the value of the inductance per unit length, for the spacing A-C. Substituting the value of Φ_{eq} from eq. (130) into eq. (127) we get

$$e_1 = L'_{12}di_2/dt + L'_{13}di_3/dt. \quad (131)$$

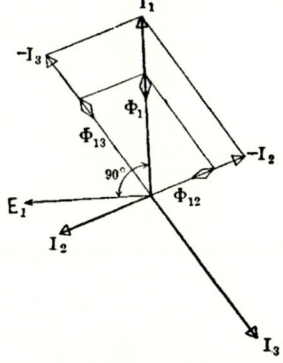

This shows that with an unequal spacing the effect of the mutual induction of the phases cannot be replaced by an equivalent inductance in each phase, because, generally speaking, the currents i_2 and i_3 cannot be eliminated from this equation by means of eq. (126).

Fig. 48.—The currents and fluxes in a three-phase line with a *symmetrical* spacing.

Let us apply now eq. (131) to the case of sinusoidal currents and voltages. Let the current in the wire B be $i_2 = \sqrt{2}I_2 \sin 2\pi ft$, where I_2 is the effective value of the current; then the first term on the right-hand side of eq. (131) becomes $2\pi fL'_{12}\sqrt{2}I_2 \cos 2\pi ft$.

In the symbolic notation this is represented as $jx_{12}'I$, where x_{12}' $=2\pi f L_{12}'$ is the reactance corresponding to L_{12}', I_2 is the vector of the effective value of the current in the wire B, and j signifies that the vector $x_{12}'I_2$ is in leading quadrature with the vector I_2. Consequently, the voltage drop E_1 in the wire A, equal and opposite to the induced e.m.f., is

$$E_1 = -jx_{12}'I_2 - jx_{13}'I_3. \quad \ldots \ldots \quad (132)$$

When the currents are given, I_2 and I_3 can be expressed in the usual way through their components, and the drop E_1 is then expressed through its components as $e_1 + je'_1$. The reactances x_{12}' and x_{13}' are taken from the available tables, for the specified frequency and the appropriate spacings, or else they can be calculated using the value of L' from eq. (125).

The voltage drop E_1 in eq. (132) can be represented as if it were due to an equivalent reactance x_1' and an equivalent resistance r_1' in the phase A (the latter in addition to the actual ohmic resistance of the wire). This is possible when I_2 and I_3 can be expressed in terms of I_1, and is especially convenient whenever the phase difference between these currents and their ratio is constant. Namely, let the current I_2 lead the current I_1 in phase by ϕ_{12} electrical degrees (Fig. 49). Then

$$I_2 = (I_2/I_1)I_1(\cos \phi_{12} + j \sin \phi_{12}), \quad \ldots \quad (133)$$

where (I_2/I_1) is the ratio of the effective values of the currents, apart from their phase relation. Multiplying the vector I_1 by (I_2/I_1) changes its magnitude to that of I_2, while multiplying it by $(\cos \phi_{12} + j \sin \phi_{12})$ turns it counter-clockwise by ϕ_{12} degrees. By analogy we also have that

$$I_3 = (I_3/I_1)I_1(\cos \phi_{13} + j \sin \phi_{13}). \quad \ldots \quad (134)$$

Both ϕ_{12} and ϕ_{13} are measured counter-clockwise. Substituting these values into eq. (132) and separating the real from the imaginary part we get

$$E_1 = I_1[(I_2/I_1)x_{12}' \sin \phi_{12} + (I_3/I_1)x_{13}' \sin \phi_{13}]$$
$$-jI_1[(I_2/I_1)x_{12}' \cos \phi_{12} + (I_3/I_1)x_{13}' \cos \phi_{13}]. \quad \ldots \quad (135)$$

Thus, the drop E_1 is the same as if it were caused by a fictitious reactance

$$x_1' = -(I_2/I_1)x_{12}' \cos \phi_{12} - (I_3/I_1)x_{13}' \cos \phi_{13}, \quad (136)$$

and a fictitious resistance

$$r_1' = (I_2/I_1)x_{12}' \sin \phi_{12} + (I_3/I_1)x_{13}' \sin \phi_{13}. \quad . \quad (137)$$

Both x_1' and r_1' may be either positive or negative, depending upon the constants of the circuit and of the load. The resistance

r_1' does not involve any loss of power, converted into heat; it merely shows that energy is transferred inductively from phase A into B or C, at a rate $I_1^2 r_1'$, due to a lack of symmetry in the resultant field.

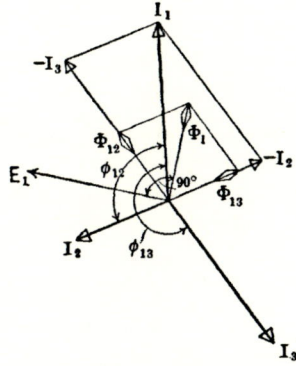

Prob. 27. Show that with a balanced three-phase load

$$\left. \begin{array}{l} x_1' = 0.5(x_{12}' + x_{13}'); \\ r_1' = 0.866(x_{12}' - x_3')_{.1} \end{array} \right\} \quad . \quad (138)$$

Fig. 49.—The currents and fluxes in a three-phase line with an *unsymmetrical* spacing.

Prob. 28. When the three wires are in the same plane, the spacings being equal, and the three-phase load balanced, show that the equivalent reactance of each outside wire

$$x_o' = x_m' + 0.435f \times 10^{-3} \text{ ohm/km.,} \quad . \quad . \quad . \quad (139)$$

where x_m' is the reactance of the middle wire per kilometer, in ohms. The equivalent resistance of the middle wire is zero, and that of the two outside wires is

$$r_o' = \pm 0.753f \times 10^{-3} \text{ ohm/km.,} \quad . \quad . \quad . \quad . \quad (140)$$

where the sign *plus* refers to the wire in which the current *leads* that in the middle wire.

Prob. 29. Compare the vector diagram in Fig. 49 with that in Fig. 48, and shown that with an unsymmetrical spacing the induced voltage E_1 is not in quadrature with the corresponding current I_1, so that the action of the other two wires cannot be replaced by an equivalent inductance alone, but only by an inductance and a resistance. Show graphically that the latter may be either positive or negative.

CHAPTER XII

THE INDUCTANCE OF THE WINDINGS OF ELECTRICAL MACHINERY.

64. The Inductance of Transformer Windings. When a transformer is operated at no load, i.e., with its secondary circuit open, practically the whole flux is concentrated within the iron core. When, however, the transformer is loaded, so that considerable currents flow in both windings, appreciable leakage fluxes are formed (Fig. 50), which are linked partly with the primary winding, and partly with the secondary winding. When the load current is considerable, the primary and the secondary ampere-turns are large as compared to the exciting ampere-turns, so that at each instant the secondary ampere-turns are practically equal and opposite to the primary ampere-turns. An inspection of Fig. 50 will show that the m.m.f. acting upon the useful path in the iron is equal to the difference between the m.m.fs. of the primary and secondary windings, while the m.m.f. acting upon leakage paths is equal to the sum of the m.m.fs. of both windings.

Take, for instance, the line of force $fghk$; with respect to the part fg of its path, the secondary coil S_1 and the adjacent half P_1 of the primary coil form together a fictitious annular coil (leakage coil). The m.m.f. of this coil is equal to $\frac{1}{2}n_1 i_1$, where i_1 is the primary current, and n_1 is the number of turns in the whole primary coil P. Similarly, the coil S_2 and the part P_2 of the primary coil may be said to form another fictitious coil linking with the part hk of the path of the lines of force.

It will be seen from the dots and crosses that the m.m.fs. of the two fictitious coils assist each other, and that the paths of the leakage flux are as indicated by the arrow-heads. Some lines of force are linked with the total m.m.f. of the fictitious coils, others are linked with only part of the turns. Although

208

the reluctance of the leakage paths is very high as compared to that of the useful path in iron, yet the m.m.f. acting upon

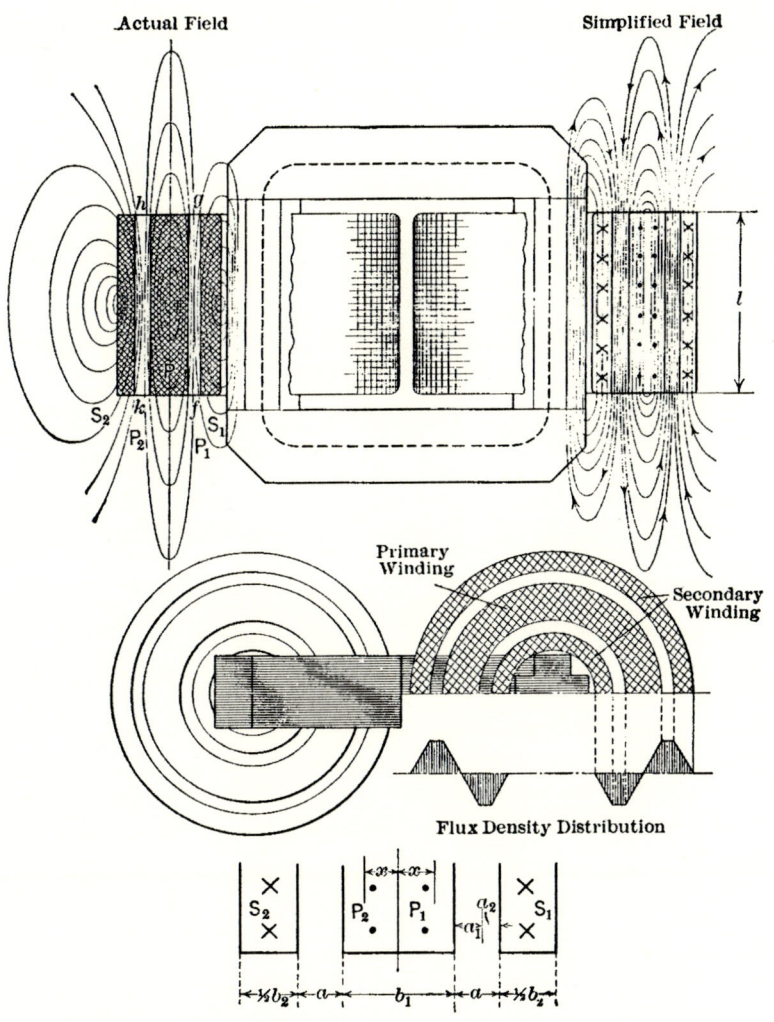

Actual Field

Simplified Field

Primary
Winding

Secondary
Winding

Flux Density Distribution

Fig. 50.—The leakage field in a transformer with cylindrical coils.

the leakage paths is also many times greater than that acting upon the path in iron. As a consequence, the leakage fluxes reach appreciable magnitudes.

The leakage fluxes induce e.m.fs. in the windings in lagging quadrature with the respective currents, and thus affect the voltage regulation of the transformer. That part of the applied voltage which balances these e.m.fs. is known as the reactance drop in the transformer. It is customary to speak about the primary reactance and the secondary reactance, also about the primary and the secondary leakage fluxes, as if they had a separate and independent existence. However, it must be understood that each leakage flux is produced by the combined action of both windings, as is explained above. Moreover, where the leakage fluxes enter the iron they combine with the main flux in the proper direction, so that they form there only a component of the resultant flux.

In reality, the primary ampere-turns are not exactly equal and opposite to the secondary ampere-turns, so that, in addition to the leakage fluxes shown in Fig. 50, there is a leakage flux due to the magnetizing ampere-turns. However, this correction is negligible, when the load is considerable, and the calculation of the leakage flux is greatly simplified by assuming the primary ampere-turns to be exactly equal and opposite to the secondary ampere-turns.

The effect of the leakage reactance upon the performance of a transformer is treated in *The Electric Circuit;* there the value of the reactance is supposed to be given. Here the problem is to show how to calculate the leakage inductance from the given dimensions of a transformer. The two types of winding to be considered are the one with cylindrical coils (Fig. 50) and the one with flat coils (Fig.51). Both types of winding can be used with any of the three kinds of magnetic circuit which are used with transformers (Figs. 12, 13, and 14).

The problem of calculating the leakage inductance, according to the fundamental formula (106), is reduced to that of finding the permeances of the individual paths of the leakage flux. It would be out of the question here to determine the actual paths and to express their permeances mathematically. Therefore, in accordance with Dr. Kapp's proposal.[1] simplified paths are assumed, shown in Fig. 50 to the right. The inductance so calculated is corrected by an empirical factor, obtained from experiments on transformers of similar type and proportions.

[1] G. Kapp, *Transformers* (1908), p. 177.

The simplifying assumptions are (a) that the paths within and between the coils are straight lines, and (b) that the reluctance of the paths in the space outside the coils is negligible, because the cross-section of these paths is practically unlimited.

(1) *Cylindrical Coils.* We shall calculate first the primary inductance of a transformer having cylindrical coils, i.e., the inductance due to the linkages of the leakage flux with the primary winding. The permeance of the path of the complete linkages is $\mathcal{P}_{c1} = \mu . a_1 O_m / 2l$ perms, where O_m is the mean length of a turn in the coil P, and a_1 is the radial thickness of the flux. The notation is shown in the detail drawing, at the bottom of Fig. 50. In this expression $a_1 O_m$ is the mean cross-section of the path, being an average between the cross-sections within the spaces P_1–S_1 and P_2–S_2. The length of the paths within the coils is $2l$, and the reluctance of the paths outside the coils is neglected. This path is linked with n_1 turns.

Similarly, the permeance of an infinitesimal annular path within the primary coil, at a distance x from its center, and of a width dx, is $d\mathcal{P}_{p1} = \mu O_m dx / 2l$ perms. This path is linked with $n_{p1} = n_1 (2x/b_1)$ turns. Substituting these values into eq. (106) we obtain

$$L_1 = (\mu n_1{}^2 O_m / 2l) \left[a_1 + \int_0^{\frac{1}{2}b_1} (2x/b_1)^2 dx \right],$$

or

$$L_1 = (\mu n_1{}^2 O_m / 2l)(a_1 + \tfrac{1}{6}b_1) \text{ perms.} \quad . \quad . \quad . \quad (141)$$

By a somewhat similar reasoning we should find for the combination of the two secondary coils, assuming them to be connected electrically in series,

$$L_2 = (\mu n_2{}^2 O_m / 2l)(a_2 + \tfrac{1}{6}b_2) \text{ perms.} \quad . \quad . \quad . \quad (142)$$

In the operation of a transformer it is the total equivalent inductance of the two windings reduced to one of the circuits that is of importance. Since resistances and reactances can be transferred from the primary circuit to the secondary or *vice versa*, when multiplied by the square of the ratio of the numbers of turns (Art. 44b), the equivalent inductance, reduced to the primary circuit, and *per leg* of the core, is

$$\begin{aligned} L_{eq} &= L_1 + (n_1/n_2)^2 L_2 \\ &= k(\mu n_1{}^2 O_m / 2l) \, [a + \tfrac{1}{6}(b_1 + b_2)]10^{-8} \text{ henrys.} \quad . \quad (143) \end{aligned}$$

All the lengths here are expressed in centimeters, and $\mu = 1.257$, also $a = a_1 + a_2$ is the spacing between the coils, which is a known quantity. Thus, the unknown distances a_1 and a_2 which enter into the expressions for L_1 and L_2 are eliminated from the formula for the equivalent inductance.

The coefficient k corrects for the difference between the actual linkages shown in Fig. 50 at the left, and the assumed linkages shown at the right. The values of k, found from experiments, vary within quite wide limits, depending upon the proportions of the coils. For good modern transformers Arnold gives the limits of k between 0.95 and 1.05.[1] See also eq. (147) below. Formula (143) gives the inductance of one leg only; the equivalent inductance of the whole transformer depends upon the electrical connections between the coils.

In designing a transformer the coils are usually arranged in such a way as to reduce the leakage reactance to the least possible amount.[2] Eq. (143) shows that in order to achieve this result, a comparatively small number of turns must be used, and the coils must be thin and long. The space a between the coils must be kept as small as is compatible with the requirements for insulation and cooling.

The usual arrangement of coils shown in Fig. 50 gives a considerably smaller leakage inductance than the simpler arrangement shown in Fig. 12. Namely, with the arrangement shown in Fig. 12, the permeance of the path of the complete linkages in the space between the coils is $\mu a_1 O_m / l$. This expression differs from that used before in that l stands in the denominator in place of $2l$. For the partial linkages $n_p = n_1 (x/b_1)$, where x is measured now from the edge of the primary coil, furthest from the secondary coil. Thus, the primary inductance is in this case

$$L_1 = (\mu n_1{}^2 O_m / l) \left[a_1 + \int_0^{b_1} (x/b_1)^2 dx \right],$$

or

$$L_1 = (\mu n_1{}^2 O_m / l)(a_1 + \tfrac{1}{3} b_1).$$

By symmetry we can write the expression for L_2, and hence,

[1] E. Arnold *Wechselstromtechnik* (2d edition), Vol. 1., p. 561.
[2] In some cases a considerable leakage reactance is specified as a protection against violent short circuits.

after combining,

$$L_{eq} = k(\mu n_1{}^2 O_m/l)[a + \tfrac{1}{3}(b_1 + b_2)]10^{-8}, \quad \cdots \quad (144)$$

This value is between two and four times as large as the value given by eq. (143). For this reason, in most transformers, the low-tension coil is split into two sections; compare also with Fig. 14.

Formula (143) and the values of k given above have been deduced for the core-type transformer. It is clear, however, that the same formulæ will apply to the shell-type and the cruciform type transformers with cylindrical coils, though the coefficient k may have different values in each case. Until more reliable and numerous experimental data are available the same values of k will have to be used for these types as for the core-type.[1]

(2) *Flat Coils.* With flat coils (Fig. 51) the inductance of a part of the winding between AB and CD can be calculated in precisely the same way as in Fig. 50. If the primary winding is split into q sections, the inductance per section, by analogy with eq. (143), is

$$L_{eq}/q = k[\mu(n_1/q)^2 O_m/2l][a + \tfrac{1}{6}(b_1 + b_2)]10^{-8} \text{ henrys}, \quad (145)$$

where the dimension l is again measured in the direction of the lines of force and O_m is the mean length of a turn. The dimensions a, b_1, and b_2 are indicated in Fig. 51. The inductance of the whole winding is

$$L_{eq} = k(\mu n_1{}^2 O_m/2ql)[a + \tfrac{1}{6}(b_1 + b_2)]10^{-8} \text{ henrys}, \quad \cdots \quad (146)$$

where all the lengths are expressed in centimeters, and $\mu = 1.257$.

This formula presupposes that the m.m.fs. are balanced, or in other words, that there are two half-sections of the same winding at the ends; such is usually the case in order to reduce the leakage reactance. (See also Fig. 13.)

Eq. (146) shows that the leakage reactance is considerably reduced, and consequently the voltage regulation improved, by subdividing the windings and placing the primary and the

---

[1] See also the *Standard Handbook for Electrical Engineers* under Transformer, leakage reactance.

secondary windings on the core alternately. At a given voltage, and with a given type of construction, the spacing a between the coils may be considered as constant and independent of the number of sections. In transformers for extra-high voltages a is large as compared to $\frac{1}{6}(b_1 + b_2)$, so that the leakage reactance is almost inversely proportional to the number of sections q. In low-voltage transformers a is small as compared to b_1 and b_2; hence, L_{eq} is almost inversely proportional to q^2, because b_1 and b_2 are themselves inversely proportional to q. Thus,

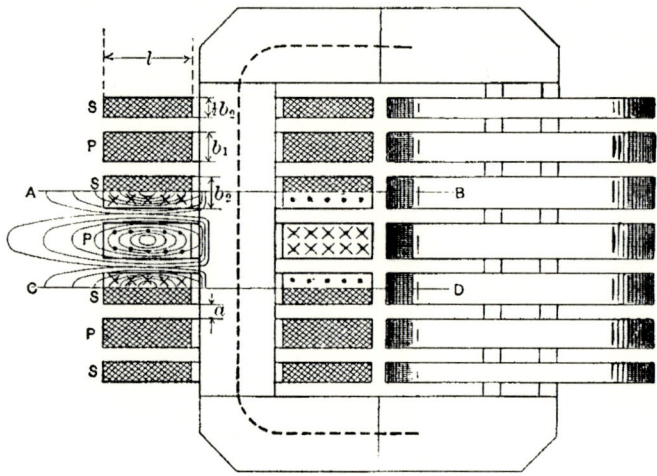

FIG. 51.—The leakage field in a transformer with flat coils.

in general, the inductance of a transformer is inversely proportional to q^n, where n has a value between 1 and 2.[1]

Dr. W. Rogowski has given an exact mathematical solution for the flux distribution in the case of flat transformer coils, assuming the coils and the core to be indefinitely long in the direction perpendicular to the cross-section shown in Fig. 51.[2]

[1] For experimental data in regard to the effect of the subdivision and arrangement of transformer windings upon the leakage reactance see Dr. W. Rogowski, *Mitteilungen Ueber Forschungsarbeiten*, Heft 71 (Springer, 1909), p. 18, also his article Ueber die Streuung des Transformators, *Elektrotechnische Zeitschrift*, Vol. 31 (1910), pp. 1035 and 1069; also Faccioli, Reactance of Shell-type Transformers, *Electrical World*, Vol. 55 (1910), p. 941.

[2] Dr. W. Rogowski, *loc. cit.*

With certain simplifying assumptions he arrived at the same formula for inductance as eq. (146) in which approximately

$$k = 1 - (b_1 + b_2 + 2a)/(2\pi l).$$

Because of the simplifying assumptions made in the deduction of this formula, the values of k calculated from the results of tests on actual transformers differ slightly from those given by the formula. Let k' be an empirical correction coefficient, then

$$k = k'[1 - (b_1 + b_2 + 2a)/(2\pi l)]. \quad . \quad . \quad . \quad . \quad (147)$$

In Dr. Rogowski's experiments the actually measured inductance was on the average 6 per cent higher than the calculated one. Until more experimental data are available it is therefore advisable to use in eq. (147) the value of $k' = 1.06$. Eq. (147) is applicable to transformers of all the three types (Figs. 12 to 14), though in the case of a shell-type or cruciform transformer, the presence of iron outside the coils tends to increase the value of k'. However, Dr. Rogowski states, that with the space usually allowed for insulation between the coils and the iron, the influence of the iron in increasing the leakage reactance is negligible. Eq. (147) holds approximately true for cylindrical coils also, though there are as yet no conclusive tests for the value of the correction factor to be used with such coils.

The general similarity between the equations for leakage inductance given above raises the question, as to what element they possess in common. This is found in the conception of a leakage coil, which is the " fictitious coil " spoken of above. An inspection of Figs. 50 and 51 will show that the successive lines of force converge upon lines which may be called the " hearts " of the flux system. These hearts are located in places where the net m.m.f. is zero. This is at the edge of the half-coils and at the center of the whole-coils, in the two figures mentioned. In deriving eq. (144) for the case where the coils are not split, the heart is assumed to be at the edge of both coils. If we define a leakage coil as that part of the winding between two successive hearts, then eq. (144) will always apply to it. In eq. (143) b_1 and b_2 refer to the width of the double leakage coil, hence if we substitute in eq. (144) $\frac{1}{2}b_1$ and $\frac{1}{2}b_2$ for b_1 and b_2, we get the coefficient $\frac{1}{6}$, which appears in eq. (143). In eq. (143) n_1 and L_{eq} refer to the double leakage coil. Substituting in eq. (144) $\frac{1}{2}n_1$ for n_1 and $\frac{1}{2} L_{eq}$ for L_{eq}, we get eq. (143). Thus, the differences between the two equations is readily explained. In case the coils are divided in any unusual manner, we must first locate the hearts by noticing where the m.m.fs. are balanced. Then we should figure out the inductance by eq. (144) for each leakage coil separately. The only precaution to be observed is that the various quantities refer to the leakage coil. Finally (if the coils are in series) we should add the various inductances together. The arrangement with half coils on the ends gives the minimum of inductance for a given number of coils.

Prob. 1. The approximate assumed dimensions of a 15-kva., 2200/110 v., 60-cycle, cruciform-type transformer with cylindrical coils (Fig. 14) are: $O_m = 140$ cm.; $b_1 = 4.5$ cm.; $b_2 = 3$ cm.; $a = 1$ cm. The maximum useful flux is 1.03 megalines. Show that the relationship between the height l of the winding and the percentage reactive voltage drop is $xl = 164$. Assume $k = 1.10$.

Prob. 2. Referring to the preceding problem, what is the permeance of the space between the outside low-tension coil and high-tension coil, per centimeter of the height of the coils, and what is the effective value of the flux density in this space, at full load?
                  Ans.   197.5 perms per cm., $3420/l$ lines per sq. cm.

Prob. 3. Each leg of a core-type transformer is provided with six flat high-torsion coils of 530 turns each, interposed with the same number of low-tension coils of 40 turns each, one of the low-tension coils being split in two and placed at the ends. The high-tension coils are wound of 3 mm. round wire, 53 layers, 10 turns per layer ($b_1 = 3$ cm.); the low-tension coils are wound of 8 mm. square wire, in 20 layers, 2 turns per layer ($b_2 = 1.6$ cm.). The distance between the coils is 20 mm. Taking the inductance of this transformer to be unity, calculate the relative inductances of the transformer when the high-tension winding is divided into three coils and also into two coils, assuming k, l, and a to be the same in all cases.   Ans.   2.55; 4.66.

Prob. 4. Solve the preceding problem, taking into account the change in k.   Ans.   2.42; 4.14.

Prob. 5. The following results were obtained from a short-circuit test on a 22/2-kv., 2500-kva., 60-cycle, shell-type transformer, with flat coils: With the high-tension winding short-circuited, and full rated current flowing through the low-tension winding the voltage across the secondary terminals was 73.5 v., and the wattmeter reading was 27 kw. The transformer winding consists of 12 high-tension coils of 100 turns each, and of 11 low-tension coils interposed between the high-tension coils, together with 2 half-coils at the ends. The dimensions of the coils are: $O_m = 2.6$ m.; $l = 18$ cm.; $b_1 = 16$ mm.; $b_2 = 10$ mm.;

$a = 12$ mm. Calculate the correction factor k' in formula (147). Hint: Eliminate the ir drop, using the wattmeter reading. Ans. About 1.06.

Prob. 6. What is the greatest permissible thickness of the coils of a 60-cycle transformer, if the reactive drop must not be larger than three times the resistance drop? The ducts a are 1 cm. The space factor of the copper in each coil is 0.55. The primary coils and the whole coils of the secondary are of the same thickness. $k = 0.98$. Hint: Assume O_m, l, and n_1 and show that they cancel out.

Ans. $b = 2.3$ cm.

Prob. 7. In order to provide a better cooling, and at the same time save on insulation, two flat high-tension coils are frequently placed side by side, with a small air-space in between, and in the same way the low-tension coils may be subdivided. Show that no leakage flux passes in the space between the two adjacent coils which belong to the same winding, so that the inductance of the winding is not appreciably increased by these spaces, and can be calculated as if these spaces did not exist.[1]

Prob. 8. Compare the formulæ given above and the numerical values of transformer leakage reactance obtained therefrom with the formulæ and data given in the *Standard Handbook for Electrical Engineers*. Do this for any transformer, the dimensions of which are available.

Prob. 9. The equivalent reactance of a transformer is x_1 reduced to the primary circuit, and is x_2 reduced to the secondary circuit. All the primary coils are connected in series, and all the secondary coils are also in series. Show that these equivalent reactances become x_1/c_1^2 and x_2/c_2^2 respectively, when the primary winding is divided into c_1 branches in parallel, and the secondary winding is divided into c_2 branches in parallel. The division is supposed to be made in such a way as to keep the m.m.fs. in the adjacent coils balanced. Hint: If the equivalent reactance of one primary branch is x_1', that of c_1 branches in series is $x_1 = x_1 c_1'$, and that of c_1 branches in parallel is x_1'/c_1 no matter whether the secondary coils are connected in series or in parallel.

Prob. 10. Prove that the equivalent reactance of a transformer is the same, whether the coils are in series or in parallel, provided that the total number of turns in series is the same.

Note: Sometimes, because of the difficulty in using heavy conductors, it is desirable to multiple the coils. In such case, the parallel coils *must* have the same number of turns, and they should be symmetrically arranged, so as to prevent exchange currents.

65. The Equivalent Leakage Permeance of Armature Windings.

In certain problems relating to the design and operation of electrical machinery it is necessary to calculate the inductance

---

[1] This fact has been verified experimentally; see Arnold, *Wechselstromtechnik*, Vol. 2 (1910), p. 29.

of the armature windings. This inductance affects the performance of the machine, because the leakage fluxes created by the armature currents induce e.m.fs. in the machine. Such leakage fluxes are shown in Figs. 23 and 36, in an induction machine and a synchronous machine respectively. For purposes of theory and computation these leakage fluxes are usually subdivided into three parts, namely:

(a) Leakage fluxes linked with the parts of the windings embedded in the armature iron (Figs. 36 and 54). These paths are closed partly through the slots, and partly through the tooth-tips (slot leakage and tooth-tip leakage).

(b) Leakage fluxes linked with the parts of the armature windings in the air-ducts.

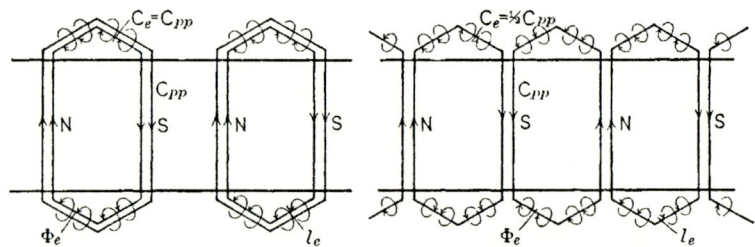

FIG. 52.—Undivided end-connections.   FIG. 53.—Divided end-connections.
$\alpha = 1.$ $\alpha = \frac{1}{2}.$

(c) Leakage fluxes linked with the end-connections of the armature windings (Figs. 52 and 53).

Usually the fluxes (a) and (b) are merely distortional components of the main flux of the machine, and only the fluxes (c) have a real existence.

It will be readily seen that the paths of the tooth-tip leakage and of that around the end-connections are too complicated to allow the corresponding permeances to be calculated theoretically. For this reason, various empirical and semi-empirical formulæ are used in practice for estimating the leakage inductance of armature windings, the coefficients in these formulæ being determined from tests on similar machines.

The most rational procedure is to express the inductance through the equivalent permeance of the paths, as defined by eq. (106a) in Art. 58. Let there be C_{pp} conductors per pole per phase, such as are indicated for instance in Fig. 15.   Then the inductance of a machine, per pole per phase, is given by the equation

$$L_{pp} = C_{pp}{}^2 \mathcal{P}_{eq}, \quad \ldots \ldots \ldots \quad (148)$$

where $\mathcal{P}_{eq}$ is an empirical value of the equivalent permeance per pole per phase.   This formula presupposes that all the partial linkages are replaced by the equivalent complete linkages embracing all the C_{pp} conductors.   Moreover, the value of $\mathcal{P}_{eq}$ is such as to take into account the inductive action of the other phases upon the phase under consideration.   The total inductance of the machine, per phase, depends upon the electrical connections in the armature winding.   If all the p poles are connected in series, the foregoing expression for L_{pp} must be multiplied by p; if there are two branches in parallel, the inductance of each branch is $\frac{1}{2}pL_{pp}$, and the combined inductance of the whole machine is $\frac{1}{2}(\frac{1}{2}pL_{pp}) = \frac{1}{4}pL_{pp}$.

The leakage inductance of a machine is usually determined by sending through it an alternating current of a known frequency, under conditions which depend upon the kind of the machine (the field to be removed in a synchronous machine, and the armature to be locked in an induction machine).   From the readings of the current of the applied voltage and the watts input, the reactance x of the machine is calculated (after eliminating the ohmic drop).   Then, knowing the frequency f of the supply and the number of poles of the machine, the inductance $L_{pp} = x/(2\pi f p)$ per pole is calculated.   Substituting into eq. (148) this value of L_{pp} and the known number of conductors C_{pp}, the equivalent permeance $\mathcal{P}_{eq}$ per pole per phase is determined.   By performing such tests on machines built on the same punching, but of different embedded lengths, the permeance due to the embedded parts of the winding is separated from that due to the end-connections; the values so obtained are then used in new designs.

The leakage permeances in the embedded parts are proportional to the widths of these parts in the direction parallel to the shaft, in other words, to the length of conductors which are surrounded by the leakage lines.   Experiment shows that the permeance of the paths in the air-ducts and around the end-connections is also approximately proportional to the lengths of the corresponding parts of the armature coils. Since all these permeances are in parallel, $\mathcal{P}_{eq}$ is equal to their sum, or

$$\mathcal{P}_{eq} = \mathcal{P}_i' l_i + \mathcal{P}_a' l_a + \alpha \mathcal{P}_e' l_e \quad \ldots \quad \text{(149)}$$

Here the letter l denotes the lengths in cm. of the coil, or, what is the same thing, the width of the paths of flux. The subscripts i, a, and e refer to the iron, the air-ducts, and the end-connections respectively. Thus, l_i is the *semi-net* length of the core, that is, the length exclusive of ducts but inclusive of the space between the laminations.[1] The corresponding permeance per centimeter is $\mathcal{P}_i'$. The sign "prime" signifies that the corresponding $\mathcal{P}$'s refer to one centimeter width of path.

The coefficient α is equal to 1 when the end-connections are arranged as in Fig. 52, and $\alpha = \frac{1}{2}$ when they are arranged according to Fig. 53. Namely, in the first case the number of conductors C_e per group of the end-connections is equal to the number C_{pp} in the embedded part. In the second case C_e is equal to $\frac{1}{2}C_{pp}$. In the first case there are as many groups of end-connections per phase as there are poles; in the second case there are twice as many groups per phase as there are poles, so that two groups (one pointing to the right and one to the left) must be counted per pole. Thus, with undivided end-connections, the inductance is $C_{pp}{}^2 \mathcal{P}_e' l_e$, while with divided end-connections it is $(\frac{1}{2}C_{pp})^2 \mathcal{P}_e' . 2l_e = C_{pp}{}^2 . \frac{1}{2}\mathcal{P}_e' l_e$. This accounts for the value of $\alpha = \frac{1}{2}$, in formula (149), and shows that the inductance due to the end-connections is reduced twice by subdividing them into two groups. When estimating the leakage reactance it is therefore necessary to know the exact arrangement of the end-connections.

In preliminary calculations, before the armature coils are drawn to scale, the length l_e of the end-connections in a full-pitch winding is usually assumed to be equal to about 1.5τ, where τ is the pole pitch. For fractional-pitch windings, l_e varies roughly as the winding-pitch, or $l_e = 1.5\gamma\tau$ (see Art. 29).

---

[1] The semi-net length is used in getting the leakage permeance in the slot, because the flux spreads as it comes out of the iron almost immediately. The spaces between the laminations do not affect the density in the air because they are so small as compared to the dimensions of the slot.

Substituting the value of $\mathcal{P}_{eq}$ from eq. (149) into eq. (148) we obtain

$$L_{pp} = C_{pp}^2 (\mathcal{P}_i' l_i + \mathcal{P}_a' l_a + \alpha \mathcal{P}_e' l_e) 10^{-8} \text{ henrys.} \quad . \quad (150)$$

In the following three articles formula (150) is applied to the calculation of the leakage reactance of

(a) Induction machines;
(b) Synchronous machines;
(c) Coils undergoing commutation in a direct-current machine.

In each case somewhat different values of the unit permeances $\mathcal{P}'$ are used, because of the diversity of the magnetic paths.

66. The Leakage Reactance in Induction Machines. It is explained in Art. 35 and shown in Fig. 23 that the actual flux in a loaded induction machine is the resultant of three fluxes, of which the useful flux $\mathit{\Phi}$ is linked with both the primary and the secondary windings. The component fluxes $\mathit{\Phi}_1$ and $\mathit{\Phi}_2$, linked with the primary and secondary windings respectively, are called the *leakage fluxes*. They induce in the windings e.m.fs. in quadrature with the corresponding currents, and these e.m.fs. have to be balanced by a part of the terminal voltage. Consequently, that part of the applied voltage which is balanced by the useful flux is reduced; in other words, the useful flux and the useful torque are reduced with a given current input. As a matter of fact, the maximum torque and the overload capacity of an induction machine are essentially determined by its leakage fluxes, or what amounts to the same thing, its leakage inductances.

Knowing the primary and secondary leakage reactances, the actual induction machine is replaced by an equivalent electric circuit (or a circle diagram is drawn for it), after which its performance can be predicted at any desired load. The problem here is to determine the values of these leakage reactances and inductances, from the given dimensions of a machine. The rest of the problem is treated in the *Electric Circuit*.

The leakages fluxes, which are indicated schematically in Fig. 23, are shown more in detail in Fig. 54. The primary conductors in one of the phases and under one pole are marked with dots, and the adjacent secondary conductors are marked with crosses, to indicate the currents which are flowing in them. Assuming the rotor to be provided with a squirrel-cage

winding, the distribution of the secondary currents is practically an image of the primary currents. Neglecting the magnetizing ampere-turns necessary for establishing the main or
useful flux, the secondary ampere-turns per pole per phase are
equal and opposite to the primary ampere-turns. A similar
assumption is also made in the preceding article, in the case of
the transformer. This assumption is not as accurate in the
case of an induction machine, because here the magnetizing
current is proportionately much larger, due to the air-gap;
nevertheless, the assumption is sufficiently accurate for most

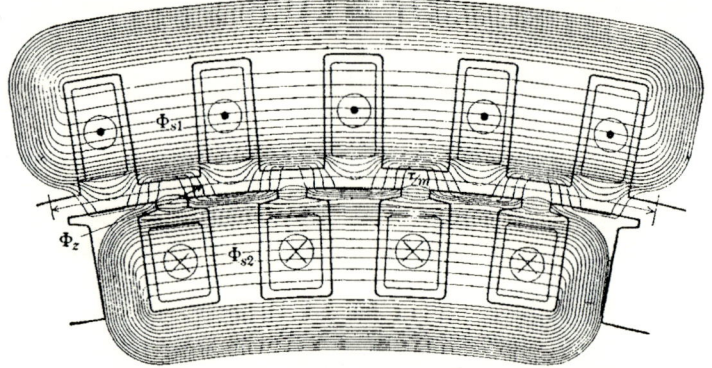

Fig. 54.—The slot and zig-zag leakage fluxes in an induction machine.

practical purposes. Even if the magnetizing current is equal
to say 25 per cent of the full-load current, the difference between
the primary and the secondary ampere-turns should be less
than 10 per cent, because the magnetizing current is considerably
out of phase with the secondary current.[1]

With this assumption, the primary and the secondary current
belts shown in Fig. 54 may be considered as two sides of a narrow
fictitious coil which excites the leakage flux, causing it to pass
circumferentially along the active layer.[2] Neglecting the mutual

[1] See the circle diagram of an induction motor, for instance, in the author's
Experimental Electrical Engineering, Vol. 2, p. 167.

[2] Although the secondary frequency is different from the primary, with
respect to the *revolving* rotor it is the same as the primary frequency with
respect to the stator. Let s be the slip expressed as a fraction of the primary
frequency. Then the speed of the rotor is $(1 - s)$, and the frequency of the
secondary currents with respect to a fixed point on the stator is $s + (1 - s) = 1$.

action of the consecutive phases, the length of this flux is approximately τ/m, where τ is the pole pitch and m is the number of the stator phases. Part of the flux is linked with the primary current belt, and part with the secondary belt. The conditions are essentially the same as between the transformer windings P_1 and S_1 in Fig. 50. Knowing the equivalent permeances of the individual paths the inductance can be calculated from eq. (150).

In an induction machine with a squirrel-cage rotor the total leakage in the embedded part may be resolved into three components shown in Fig. 54, namely:

(1) The primary slot leakage, Φ_{s1};

(2) The secondary slot leakage, Φ_{s2};

(3) The tooth-tip or zigzag leakage, Φ_z.

The fluxes Φ_{s1} and Φ_{s2} are alternating fluxes of the frequency of the corresponding currents. The zigzag flux Φ_z varies according to a much more complicated law, because the permeance of its path changes from instant to instant in accordance with the relative position of the stator and rotor teeth; compare positions (1) and (2) in Fig. 23. Moreover, Fig. 54 shows only the simplest case, which never occurs in practice, namely; when the stator tooth pitch is equal to that in the rotor. In reality, the two pitches are always selected so as to be different, in order to avoid the motor sticking at sub-synchronous speeds (due to the higher harmonics in the fluxes and in the currents) Therefore, the paths of the zigzag leakage flux are much more complicated than is shown in Fig. 54, and in calculations the average permeance of the zigzag path is used.

In a machine with a phase-wound rotor the main flux is further distorted, due to the fact that the primary and secondary phase-belts are not exactly in space opposition at all moments. While the total m.m.fs. of the primary and secondary are balanced, there is a local unbalancing which changes from instant to instant. This distortion is the same as if it were due to an additional leakage, which was named by Professor C. A. Adams the *belt* leakage.[1] This part of the leakage usually constitutes but a small part of the total leakage, and will not be considered here separately. Those interested are referred to

[1] C. A. Adams, The Leakage Reactance of Induction Motors, *Trans. Intern. Electr. Congress*, St. Louis, 1904, Vol. 1, p. 711.

the original paper and to the works mentioned at the end of this article.

Let there be C_{pp_1} conductors per pole per phase in the stator winding; then the fictitious coil (Fig. 54) made up of the primary and secondary conductors has C_{pp_1} turns, when reduced to the primary circuit. This is because the secondary winding can be replaced by an equivalent winding with a " one to one " ratio, that is, with the same number of conductors as the primary winding. In this case, the secondary current is equal to the primary current (Art. 44b). Therefore, eq. (150) can be made to give the equivalent inductance, including the primary inductance and the secondary inductance reduced to the primary circuit, provided that the permeances of the paths linking with the secondary conductors are included in the values of $\mathcal{P}''$s. Such is naturally the case when the values are determined from a test with the rotor locked.

Extended tests have shown that in a given line of machines the permeance $\mathcal{P}_i'$ is inversely proportional to the peripheral length of the equivalent leakage flux, that is, inversely proportional to (τ/m), where τ is the pole pitch and m is the number of primary phases.[1] This shows that the permeance per centimeter of peripheral length of the active layer is fairly constant, in spite of different dimensions and proportions, as long as these are varied within reasonable limits. Thus

$$\mathcal{P}_i' = \mathcal{P}_i'' \, (\tau/m).$$

where $\mathcal{P}_i''$ is the leakage permeance of the active layer in the embedded part per one centimeter of axial length and per centimeter of the peripheral length of the path. Thus, the final formula for the equivalent leakage inductance of an induction machine, per pole per phase, reduced to the primary circuit, is

$$L_{eqpp} = C_{pp_1}{}^2 [\mathcal{P}_i'' l_{i}(\tau/m) + \mathcal{P}_a' l_a + a\mathcal{P}_e' l_e] 10^{-8} \text{ henrys.} \quad (151)$$

In this formula the following average values of unit permeance may be used for machines of usual proportions, unless more accurate data are available.

[1] H. M. Hobart, *Electric Motors* (1910), table on p. 397. The values for $\mathcal{P}_i''$ given below have been computed from this table, and the results multiplied by 2, because the table gives the values of the primary permeances only.

The equivalent permeance $\mathcal{P}_i''$ of the embedded part in perms per centimeter of the semi-net axial length of the machine, and per centimeter of peripheral length of the air gap, is for

Open slots	Half-open slots	Completely closed slots
11.5	14.5	18.

The equivalent permeance around the end-connections, and around the parts of the conductors in the air-ducts, decreases with the increasing number of slots per pole per phase, for the same reason that the slot permeance decreased. In induction motors usually at least three slots are used per pole per phase, and under these conditions Mr. H. M. Hobart uses $\mathcal{P}_e' = 0.8$ perm per centimeter, with phase-wound rotors, and $\mathcal{P}_e' = 0.6$ perm per centimeter with squirrel-cage rotors. $\mathcal{P}_a'$ may be taken in all cases equal to 0.8 perm per cm. The lengths l_e and l_a are always understood to refer to the stator winding.

The foregoing data refer to machines with full-pitch windings in the stator and in the rotor. With a fractional pitch winding the equivalent leakage permeances are somewhat smaller, due to longer phase belts and to the mutual induction of the overlapping phases. Let the winding-pitch factor (Art. 29) of the primary winding be k_{w1} and that of the secondary winding k_{w2}. Then the leakage inductance of the machine, calculated for a full-pitch winding (but for $l_e = 1.5\zeta\tau$), is multiplied by $k_{w1} \cdot k_{w2}$. This is an empirical correction, which is accurate enough for ordinary practical purposes. In reality, of two machines designed for a given duty, one with a fractional pitch winding and the other not (but otherwise both alike) the first often has a higher inductance than the second. This is because more turns are required with the fractional pitch winding, if the flux densities in the iron and in the air-gap are to be the same in both cases.[1]

With the data given above the calculation of the leakage reactance of a given induction motor is quite simple, and one engaged regularly in the commercial design of induction-motors can obtain sufficiently accurate data for their design or for the

[1] For a theoretical and experimental investigation of the effect of a fractional pitch upon the leakage reactance in induction machines see C. A. Adams, Fractional-pitch Windings for Induction Motors, *Trans. Amer. Inst. Elec. Engrs.*, Vol. 26 (1907), p. 1488.

computation of their performance, provided that he adapts the numerical values of the unit permeances to each individual case, on the basis of his previous experience. Many authors have given theoretical formulæ for the leakage inductance of induction machines.[1] These formulæ, curves, and methods, while useful in accurate work, are too elaborate to be quoted here; at any rate they are of interest only to a specialist in design. Two examples of the theoretical calculation of leakage inductance are given below, in order to show the student the general method used, and thus introduce him into the literature on the subject.

(a) *A Theoretical Calculation of the Slot Leakage Permeance.* We shall calculate the equivalent permeance for the half closed slot (Fig. 55), an open slot being a special case of it. With the notation shown in sketch, and with S_{pp} slots per pole per phase, we have:

$$\mathcal{P}_c' = \mu[b_2/s_2 + b_3/s_3 + b_4'/s_4]/S_{pp}; \quad d\mathcal{P}_p' = \mu dx/(s_4 S_{pp}),$$

and $n_x = C_{pp}x/b_4$. Hence

$$\int_0^{b_4} n_x^2 d\mathcal{P}_p' = \tfrac{1}{3}\mu C_{pp}^2 b_4/s_4 S_{pp},$$

and

$$L_{spp}' = C_{pp}^2 \cdot \mu[b_2/s_2 + b_3/s_3 + (\tfrac{1}{3}b_4 + b_4')/s_4]/S_{pp}. \quad . \quad (152)$$

The equivalent permeance of *one* slot, per unit of the semi-net axial length of the machine is

$$\mathcal{P}_s' = \mu[b_2/s_2 + b_3/s_3 + (\tfrac{1}{3}b_4 + b_4')/s_4]. \quad . \quad . \quad (153)$$

This shows that the slot permeance depends only upon the proportions of the slot and not upon its absolute dimensions.

(b) *A Theoretical Calculation of the Zigzag Leakage Permeance.* The calculation of the zigzag leakage permeance is simple only

[1] C. A. Adams, *loc. cit.*; also *Trans. Amer. Inst. Elec. Engrs.*, Vol. 24 (1905), p. 338; *ibid.*, Vol. 26 (1907), p. 1488. Hobart, *Electric Motors*, (1910), Chap. xxi; Arnold, *Wechselstromtechnik*, Vol. 5, part 1 (1909), pp. 49–54; R. Goldschmidt, Appendix to his book on *The Alternating Current Commutator Motor* (1909); R. E. Hellmund, Practische Berechnung des Streuungskoeffizienten in Induktionsmotoren, *Elektrotechnische Zeitschrift*, Vol. 31 (1910), p. 1111 and 1140; W. Rogowski, Zur Streuung des Drehstrommotors, *ibid.*, pp. 356, 1292, and 1316. See also an extensive series of articles by J. Rezelman in *La Lumière Electrique*, beginning in 1909, and in the (London) *Electrician.*

when the stator tooth-pitch and the rotor tooth-pitch are equal to each other; otherwise the paths become too complicated for mathematical analysis. Since the position of the secondary teeth varies with respect to the primary teeth, the permeance of the zigzag leakage also varies, and it is necessary to take its average value over one-half of the tooth-pitch λ, that is, between the positions (1) and (2) in Fig. 23. In some intermediate position (Fig. 55), determined by the overlap y, the reluctance of the path f, per unit of semi-net axial length of the iron, is $a/(\mu y)$ rels.

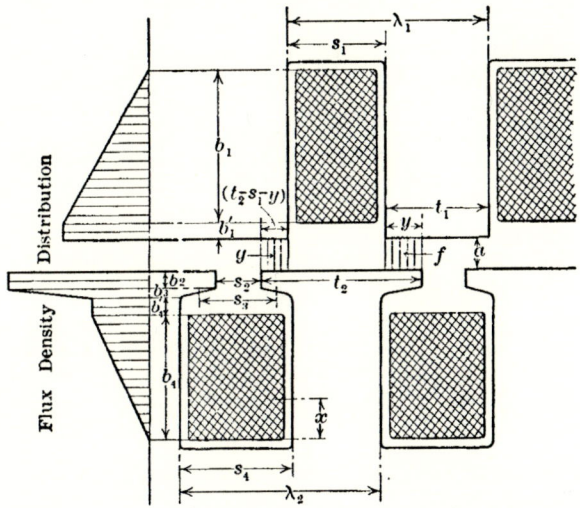

Fig. 55.—The notation used in the calculation of the slot and zigzag leakage.

The reluctance of the path g is $a/\mu(t_2 - s_1 - y)$ rels. The combined reluctance of f and g is equal to the sum of the foregoing expressions. The permeance of f and g in series, being the reciprocal of the combined reluctance, is

$$\mathcal{P}_y' = \mu y(t_2 - s_1 - y)/a(t_2 - s_1).$$

The permeance in question varies according to this law, for the positions of the secondary tooth between $y = \frac{1}{2}(t_2 - s_1)$ and $y = 0$, the tooth moving to the left. From $y = 0$ to $y = \frac{1}{2}(t_2 - s_1 - \lambda)$ the permeance is practically equal to zero, because the secondary tooth bridges over the primary slot no more. The student is advised to mark the positions of the rotor teeth on a strip of paper and to place them in different positions with respect to the primary slot in order to see the variations in the overlap. Thus, the average value of the zigzag permeance per tooth pitch is

$$\mathcal{P}_z' = \frac{1}{\frac{1}{2}\lambda}\int_0^{\frac{1}{2}(t_2 - s_1)} \mathcal{P}_y' dy = \mu(t_2 - s_1)^2 \quad (6a\lambda) \text{ perms/cm.} \quad (154)$$

Prob. 11. Draw a sketch similar to Fig. 54, but with the rotor tooth-pitch different from that in the stator, and indicate roughly the general character of the paths of the zigzag leakage.

Prob. 12. Show that increasing the number of slots per pole in a given machine does not alter materially the slot leakage, but reduces considerably the zigzag leakage.

Prob. 13. Calculate the equivalent leakage inductance per phase of a three-phase, 10-pole induction machine, with 15 slots per pole in the stator, and a phase-wound rotor. Both windings have 100 per cent pitch, the slots are semi-closed on both punchings, the individual stator coils in each phase are connected in series, and there are 20 conductors in each stator slot. The bore of the machine is 110 cm., the gross length of the core is 30 cm.; there are three vents of 8 mm. each. The end-connections are arranged according to Fig. 52.   Ans. 78.1 mh.

Prob. 14. The design of the machine in the preceding problem has been modified in the following respects: The rotor is provided with a squirrel-cage winding, the winding pitch in the stator is shortened to 80 per cent, the stator slots are made open, and the end connections are divided, as in Fig. 53. What is the inductance of the machine?

Ans. 40.3 mh.

Prob. 15. Check the values of $\mathcal{P}_i''$ given in the text above with those in Hobart's table.

Prob. 16. For the usual limits of proportions of slots, teeth, and air-gap calculate the values of $\mathcal{P}_i''$ from eqs. (153) and (154) and compare the results with the average experimental values given in this text.

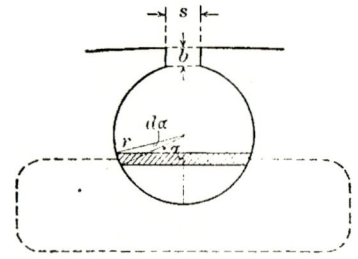
FIG. 56.—A semi-closed round slot.

Prob. 17. Calculate the equivalent leakage permeance of a round slot (Fig. 56), assuming the conductors to completely fill it, and the lines of force to be straight lines. Hint: Select the angle α as the independent variable, and integrate eq. (106) between $\alpha = 0$ and $\alpha = \pi$.

See Arnold, *Wechselstromtechnik*, Vol. 4 (1904), p. 44.

Ans. $\mathcal{P}_s' = \mu(0.623 + b/s)$ perms per cm.

67. The Leakage Reactance in Synchronous Machines. The physical nature of the armature reactance in a synchronous machine is explained in Art. 46; the influence of this reactance upon the performance of a machine is shown in Figs. 37, 38, 40, and 41. The problem here is to calculate the numerical value of this reactance for a given machine, using eq. (150) with empirical coefficients $\mathcal{P}'$.[1]

It may also be stated here that for standard machines, particularly in preliminary estimates, the ix drop is sometimes taken as a certain percentage of the rated voltage of the machine instead of estimating the inductance from formula (150). In synchronous generators the ix drop at the rated volt-ampere load varies from 5 to 10 per cent of the rated terminal voltage. In synchronous motors, where some inductance is useful, the ix drop ranges from 8 to 15 per cent of the rated voltage. For 60-cycle machines, and for machines with a comparatively large number of armature ampere-turns, values must be taken nearer the higher limit. For 25-cycle machines, and for machines with a comparatively small number of armature ampere-turns, values must be taken nearer the lower limit. A considerable error in estimating the value of ix has but little effect upon the calculated performance at unity power factor, because the vector ix is then perpendicular to e (Figs. 37, 38, 40 and 41). However, a considerable error may be introduced at lower values of the power factor if the reactive drop ix has not been estimated with a sufficient accuracy.

The values of $\mathcal{P}'$ for synchronous machines are different from those given above for induction machines, because of the absence of any secondary current-belts. Parshall and Hobart [2] give the following values for $\mathcal{P}_i'$:

---

[1] For a theoretical calculation of the coefficient $\mathcal{P}'$, see Arnold, *Wechselstromtechnik*, Vol. 4 (1904), pp. 41–52; Hawkins and Wallis, *The Dynamo*, Vol. 2 (1909), pp. 901–904. For a comparison between the calculated and actually measured values see an extended series of articles by J. Rezelman in *La Lumière Électrique*, 1909–1911, and in *The* (London) *Electrician*.

[2] *Electric Machine Design* (1906), p. 478. These values are corroborated by those obtained by Pichelmayer; see his *Dynamobau*, 1908, pp. 208 and 504. Pichelmayer's values for $\mathcal{P}_i'$ (which he denotes by ζ) are somewhat high because the end-connection leakage is not considered separately. Arnold's values, given in his *Wechselstromtechnik*, Vol. 4, p. 280, should be used with discretion, because they apply to a different formula; namely,

Uni-coil windings in open slots	3 to 6 perms per cm.	
Thoroughly distributed windings in open slots	1.5 to 3 "	"
Uni-coil windings in completely closed slots	7 to 14 "	"
Thoroughly distributed windings in completely closed slots	3 to 6 "	"

The much larger values of $\mathcal{P}_i'$ for closed slots, as compared to those with open slots, were to be expected because the bridge which closes the slot offers a path of high permeance. The lower values for windings distributed in several slots per pole per phase, as compared to uni-slot windings, are due to the fact that the partial linkages become more and more pronounced as the winding is distributed into a larger number of separate coils, and also because the length of the paths is greater. This is somewhat analogous to splitting a transmission line into two or more lines; see prob. 21 in Art. 61. The greatest reduction in the value of the inductance results when the number of slots is increased from one to two; a further subdivision is of much less importance. For instance, if the permeance $\mathcal{P}_i'$ with a uni-slot coil is 7, then dividing the same coil into two slots reduces the permeance to less than 5. On the other hand, a change from four to five slots per phase per pole would hardly reduce the equivalent permeance more than from say 3.5 to 3.4. The data in the table above give rather a wide range from which to select a value of $\mathcal{P}_i'$ for a particular machine, and the designer must exercise his judgment as to whether his machine will have a permeance nearer the upper or lower limit. This judgment comes with experience, by comparing the predicted performance of machines with that actually observed.

The values of $\mathcal{P}_a'$ and $\mathcal{P}_c'$ depend upon the number of coils per group, in other words, upon the number of slots per pole per phase. Until more accurate and detailed data are available,

he considers separately the equivalent permeance $\mathcal{P}_s$ of *each slot*, instead of the group of slots per pole per phase. Thus, his formula for the leakage inductance per pole, with our notation, is $L_{pp} = S_{pp} C_s^2 \mathcal{P}_s$, where S_{pp} is the number of slots per pole per phase, and C_s is the number of armature conductors per slot. The values of unit permeance, which he gives and denotes by λ, refer to this formula.

we shall assume $\mathcal{P}_e' = \mathcal{P}_a'$, and use the following values, based upon Mr. Hobart's experiments: [1]

Number of slots per pole per phase .	1	2	3	more than 3
$\mathcal{P}_e' = \mathcal{P}_a'$, in perms per centimeter . .	0.8	0.7	0.6	0.5

With a fractional-pitch winding the inductance is somewhat reduced (see the end of the preceding article). As an empirical correction, the inductance calculated for a full-pitch winding may be multiplied by the winding pitch factor k_w.

The leakage reactance of the armature cannot be calculated from a short-circuit test, because the short-circiut current is essentially determined by the direct armature reaction. A difference of 50 or even 100 per cent in the armature reactance would change the short-circuit current by only a few per cent. A much closer approximation is obtained from the so-called *air-characteristic*.[2] Namely, it has been found by numerous experiments that the armature inductance, when the field is revolving synchronously, is nearly equal to the armature inductance with the field completely removed, and the armature supplied with alternating currents from an external source. The air-characteristic is the relation between the current and the voltage under these conditions. Eliminating the ohmic drop, the inductance of the machine is easily calculated, and the value so found can be used in the prediction of the performance of the machine, Such an air-characteristic is easily taken in the shop or in the power house before the machine is completely assembled. From the three observed curves, namely, the no-load saturation curve, the short-circuit curve, and the air-characteristic, the performance of a synchronous machine at any load can be predicted to a considerable degree of accuracy.

Prob. 18. What is the inductance per phase of a 6-pole, 3-phase, turbo-alternator, the armature of which has the following dimensions: Bore, 1.2 m., gross length of core, 1.2 m., 20 air-ducts of 1 cm. each, 90 open slots? There are 8 conductors per slot, the winding is of the two-layer type, the winding pitch is 11/15. Assume $\mathcal{P}_i' = 2.5$ perms/cm.
Ans. 24.3 mh.

Prob. 19. The inductance of the machine specified in the preceding problem was determined experimentally (from an air-characteristic),

---

[1] *Journ. Inst. Electr. Eng.* (British), Vol. 31, pp. 192 ff.

[2] Pichelmayer, *Dynamobau* (1908), p. 207.

and compared to that measured on a similar machine, the gross length
of the core of which was 80 cm. and which was provided with 12 ducts
of 1 cm. each. The equivalent leakage permeance of the shorter machine
was found to be 30 per cent less than that of the other machine. What
are the actual values of $\mathcal{P}_i'$ and $\mathcal{P}_e'(=\mathcal{P}_a')$ for both machines?

$$\text{Ans.} \quad \mathcal{P}_i' = 2.46, \; \mathcal{P}_e' = \mathcal{P}_a' = 0.56 \text{ perm/cm.}$$

Prob. 20. An alternator has 3 slots per phase per pole, and the
equivalent permeance is $\mathcal{P}_i' = 1.8$; $\mathcal{P}_e' = 0.6$. What would be the value
of the same constants per slot?   Ans. 5.4 and 1.8.

Prob. 21. A 3-phase alternator has 4 slots per phase per pole.
If the coils were connected up for a 2-phase machine without change
what would be the ratio of the new L to the old?   Ans. 3 : 2.

68. The Reactance Voltage of Coils undergoing Commutation.

Let Fig. 57 represent a part of the armature winding
and commutator of a direct-current machine, with two adjacent
sets of brushes. During the interval of time when an armature coil, such as CD, is short-circuited by a set of brushes,
the current in the coil is reversed from its full value in one
direction to an equal value in the opposite direction. The coil
is then said to undergo commutation.

Under unfavorable conditions this reversal of current is
accompanied by sparking between one of the edges of the brushes
and the commutator. Unless a machine is provided with interpoles, its output is usually limited by this sparking at the commutator. It is of importance, therefore, to have a practical
criterion for judging the quality of commutation to be expected
of a given machine. Numerous formulæ and methods have
been proposed for the purpose; all rational formulæ contain,
as a factor, the inductance of the coils undergoing commutation,
because this inductance determines essentially the law according
to which the current is reversed with the time. For this reason,
the subject of commutation is treated in this chapter, under the
general topic of the inductance of windings. The method of
calculation of the inductance and the criterion of commutation
given below are due to Mr. H. M. Hobart.[1]

A description of the phenomenon of commutation. The phenomenon of commutation may be briefly described as follows:
Let, for the sake of explanation, the armature and the commu-

[1] See Hobart, *Elementary Principles of Continuous-Current Dynamo Design*
(1906), Chap. 4; also Parshall and Hobart, *Electric Machine Design* (1906),
pp. 171–194.

tator be assumed to be stationary, and the brushes revolving in the direction of the horizontal arrow, shown in Fig. 57. Let the machine be provided with a multiple winding (lap winding), so that there are as many armature circuits and sets of brushes as there are poles. The current through each armature branch is, therefore

$$I_1 = I/p, \quad \cdots \quad \cdots \quad \cdots \quad (155)$$

where I is the total armature current and p the number of poles.

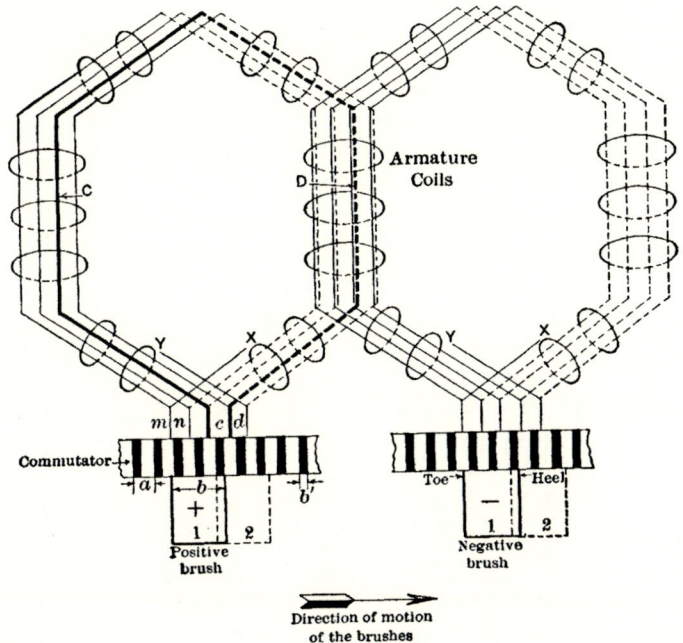

FIG. 57.—Part of the armature winding, commutator, and brushes in a direct current machine.

At each set of brushes two branches of the armature winding are connected in parallel, so that the current through each set of brushes is equal to $2I_1$ With reference to Fig. 57, it will be seen that the two armature branches, X and Y, which begin at each set of brushes, may be called, with respect to this set, the left-hand branch and the right-hand branch.

In the position of the positive brushes just preceding that marked 1, the coil CD is *not* short-circuited, and carries the full current I_1, being the first coil of the right-hand branch. The lead d is idle, and the total current $2I_1$ is delivered to the brushes through the leads c, m, and n. In the position of the positive brushes just after that marked 2, the coil CD is again *not* short-circuited, but is carrying the full current I_1 in the opposite direction, being the first coil of the left-hand branch.

In the positions of the brushes between 1 and 2 the coil CD is short-circuited by the brushes through the leads c and d, and the current in the coil changes gradually from $+I_1$ to $-I_1$. If the coil possessed no inductance, the variation in the current would be practically determined by the contact resistance between the brush and the commutator, the resistance of the coil itself and of the leads being negligible (with carbon brushes). Under these conditions the current in the short-circuited coil would vary with the time according to the straight-line law, and the current density under the heels and the toes of the brushes would be the same. This is called the " pure resistance " commutation, or the perfect commutaton, because it is not accompanied by sparking. Such a commutation is approached in machines with interpoles, when the effect of the inductance is correctly compensated for by the commutating flux (Art. 54).

In reality, the short-circuited coil possesses a considerable inductance, which has the effect of electromagnetic inertia, retarding the reversal of the current. Consequently, at the beginning of the commutation period the lead d and the corresponding commutator segment do not carry their proper share of the current, which they would carry with a perfect commutation. At the end of the commutation period the current must then be reversed quickly, because the whole current must be transferred from the lead c to the other leads. If the inductance is considerable, the current in the lead c is still of a considerable magnitude when the toe of the brush is about to leave the corresponding commutator segment. Therefore, the last period of the reversal is accomplished through the air between the brush and the segment, in the form of an electric arc. This is known as the sparking at the brushes. Besides, during the last moments of reversal, the current density under the toe is much higher than the average density under the brush, and this high density causes a glowing at the edge of the brushes, making the commutation still less satisfactory.

The average reactance e.m.f. in the coils of a full-pitch lap winding. For an empirical criterion of the quality of commutation Mr. Hobart takes *the average reactance voltage induced in the coil.* This is a reasonable criterion, because the ratio of the maximum voltage occurring when the brush leaves a segment to the average voltage, will be more or less the same in machines of usual design constants. Of course, the average reactance voltage is only a relative criterion, to be used with great discretion, and applied only for comparison with machines which proved in actual operation to commutate satisfactorily.

Let the inductance of an armature coil between two adjacent commutator segments be L_{eq}. The subscript eq (meaning equivalent) is added to indicate that the value of L includes not only the true inductance of the coil itself, but also the average inductive action of the coils which are undergoing commutation simultaneously with it. Let the frequency of the current in the coil undergoing commutation be f cycles per second. Then the current is reversed in a time $1/(2f)$. The flux changes during this time from $+L_{eq}I_1$ to $-L_{eq}I_1$. Hence, according to the fundamental eq. (26), Art. 24, the average reactance voltage, which is taken as the criterion of commutation, is

$$e_{ave} = 4fL_{eq}I_1. \quad . \quad . \quad . \quad . \quad . \quad (156)$$

In order to obtain a satisfactory commutation, the voltage e_{ave} must not exceed a certain value, determined from actual experience with machines in regular operation. Mr. Hobart recommends values for e_{ave} not to exceed 3 to 4 volts, provided that one does not depend upon the fringe flux of the main poles or upon interpoles to facilitate commutation.

The inductance L_{eq}, which enters in the foregoing formula, is calculated according to the general formula (150), as follows: Assume first that there is no common flux or mutual induction between the coil under consideration and the other coils which are simultaneously short-circuited. Then, if q is the number of turns per commutator segment (in Fig. 57 $q=1$) we must put $C_{pp}=q$. This will give the inductance of one side of the coil, say C. To obtain the inductance of both sides, C and D, the result must be multiplied by 2, or $L_{eq}=2L_{pp}$.

In reality there is a common flux which links with the coil under consideration and with the other coils undergoing commutation at the same time. This flux is changing with the time, and consequently it induces additional voltages in the coil CD. The induced e.m.f. depends upon the relative position of CD and the other coils (whether in the same slot or in the adjacent slots) and upon the rate of change of the current in each coil of the group. It would be too complicated for practical purposes to take all these factors into account with any degree of accuracy. Therefore, Hobart makes a further assumption, namely, that *the current in all the coils, which are short-circuited at the same time, varies at the same rate and that the whole leakage flux is linked with all the coils of the group* (Fig. 57).

Let s be the average number of coils simultaneously short-circuited under a set of brushes (the actual number varies from instant to instant). Consider a group of mutually influencing conductors, such as are shown at C or at D. One-half of the conductors in the same group are short-circuited by the positive brushes, the other half by the adjacent negative brushes. The total number of coils in each group is $2s$, and since by assumption the current in all of them varies at the same rate, and all of the flux is linked with all of the coils, the equivalent inductance of the coil CD is $2s$ times larger than if this coil were alone. Thus for a multiple-wound armature

$$L_{eq} = 2L_{pp}.2s = 4sq^2(\mathcal{P}_i'l_i + \mathcal{P}_a'l_a + \tfrac{1}{2}\mathcal{P}_e'l_e) \times 10^{-8} \text{ henrys.} \quad (157)$$

On the basis of Mr. Hobart's tests and until more accurate data are available, the following average values of the unit permeances may be used: $\mathcal{P}_i' = 4$ and $\mathcal{P}_a' = \mathcal{P}_e' = 0.8$ perms per centimeter.

The frequency f which enters into formula (156) is calculated as follows: The time between the positions 1 and 2 of the brushes corresponds to one-half of one cycle, because during this interval the current changes from $+I_1$ to $-I_1$. Let v be the peripheral velocity of the commutator, in meters per sec., let b be the thickness of the brushes, and b' the thickness of the mica insulation between the commutator segments, both in millimeters. The time between the positions 1 and 2 of the brushes is $(b-b')/1000v$ seconds. Hence

$$f = 500v/(b-b') \text{ cy./sec.} \quad . \quad . \quad . \quad . \quad (158)$$

The number of simultaneously short-circuited coils varies periodically with the position of the brushes. Thus, in Fig. 57 sometimes two and sometimes three coils are short-circuited by one set of brushes. On the average

$$s = (b - b')/a, \quad \cdots \cdots \quad (159)$$

where a is the width of one commutator segment including the mica insulation. Thus, all the values which enter into the formula (156) are determined, and the reactance voltage for a given machine can be easily calculated.

Formula (156) is used not only as a criterion of the commutation, but also for the calculation of the flux density, under the interpoles where such are required. Namely, this flux density must be such that the average voltage induced by the commutating flux is approximately equal and opposite to the average reactance voltage; see Art. 24. prob. 6. A still closer compensation for the influence of the inductance is achieved by properly grading the commutating flux, so as to compensate not only for the average reactance voltage, but also to some extent for the instantaneous induced e.m.fs.

The average reactance e.m.f. induced in some other direct current windings. With two-circuit wave windings two cases must be considered, namely, (a) when the machine is provided with only two sets of brushes, (b) when there are more than two sets of brushes. In the first case eq. (156) is used, where

$$I_1 = \tfrac{1}{2}I, \quad \cdots \cdots \cdots \quad (160)$$

and L_{eq} is understood to comprise the short-circuited conductors under all the poles, per commutator segment. Let there be again q turns per coil, that is, per unit of winding per pair of poles. Since the corresponding conductors under all the poles are in series, we have that $L_{eq} = pL_{pp}$. The influence of the other simultaneously short-circuited coils is expressed as before by the factor $2s$, where s is given by formula (159). Thus, for a two-circuit winding with two sets of brushes

$$L_{eq} = 2psq^2(\mathcal{P}_i'l_i + \mathcal{P}_a'l_a + \tfrac{1}{2}\mathcal{P}_e'l_e) \times 10^{-8} \text{ henrys.} \quad . \quad (161)$$

The frequency f is given as before by eq. (158).

When more than two sets of brushes are used, the sets of equal polarity are connected in parallel by the stud connections outside the armature windings, and beside there are two short-circuiting paths through the coils undergoing commutation: a long path and a short path. Thus, the problem becomes indefinite, because it is not possible to tell the relative amounts of the current through these different paths. Disregarding the short path, the criterion becomes the same as in the case of a machine with two sets of brushes only.[1] This is on the safe side, and the commutation may be expected to be better than that calculated, or at the worst, as good.

With multiplex windings, the expression for e_{ave} is the same as that given above, provided that proper values are selected for I_1, s, and f. With regard to the latter quantity it must be remembered that b' is much larger than the actual thickness of mica. Namely, with respect to the component winding under consideration the metal of the commutator segments belonging to the other component windings is equivalent to insulation. This fact must not be lost sight of in choosing the correct value for b' to be used in the expression (158).

With fractional-pitch windings the reactance voltage is smaller than with the corresponding full-pitch winding, because the conductors short-circuited under the adjacent sets of brushes (Fig. 57) are situated in part or totally in different slots, and have a smaller common magnetic flux, or none at all. When the winding pitch is reduced considerably, s instead of $2s$ must be used in the preceding formulæ; otherwise a value between s and $2s$ must be chosen, according to one's judgment.[2]

Prob. 22.   The armature of a 6-pole, 600-r.p.m., multiple-wound, direct-current machine has the following dimensions: Diameter, 85 cm.; gross length, 22 cm.; three air-ducts, 1 cm. each; 1008 face conductors.

[1] For an analysis of this case see C. A. Adams, Reactance E.M.F. and the Design of Commutating Machines, *Electrical World and Engineer*, Vol. 46 (1905), p. 346.

[2] For an advanced and more scientific theory of commutation, see Arnold, *Die Gleichstrommaschine*, Vol. 1 (1906), pp. 354 to 513; in particular the approximate formula (170) on bottom of p. 498; also Vol. 2 (1907), chapter 14. A simpler and more concise treatment will be found in Tomälen's *Electrical Engineering*. A good practical treatment will also be found in Pichelmayer's *Dynamobau*, pp. 86–118; it is considerably simplified as compared to Arnold's treatment, and is accurate enough for practical purposes, because the numerical values of unit permeances are known only approximately.

The commutator diameter is 52 cm.; the number of segments, 252; the mica insulation is 1 mm. thick; the brushes are 15 mm. thick. What is the average reactance voltage when the total armature current is 320 amp.?   Ans.   4.52 volts.

Prob. 23. Show that the answer to the preceding problem would be nine times larger if by mistake the winding were assumed to be of the two-circuit type.

Prob. 24. The peripheral velocity of a commutator is 18 meters per sec., the width of each segment (without mica) is 4.5 mm.; the thickness of the mica is 0.9 mm. The commutator is to be used in connection with a duplex winding. What is the smallest permissible thickness of the brushes if the frequency of commutation must not exceed 800 cycles per sec.?   Ans.   17.5 mm.

Prob. 25. For a perfect commutation and for an imperfect one draw the following curves to time as abscissæ: (a) the current in the short-circuited coil; (b) the currents in the leads c and d; (c) the current densities under the heel and toe of the brush. Take the width of the brushes to be equal to that of one commutator segment, and assume the mica insulation to be of a negligible thickness.

Prob. 26. Show that the width of the brushes has comparatively little net effect upon the commutation of a machine.

Prob. 27. It is desired to compensate, by means of interpoles, the reactance voltage in the machine specified in prob. 22. Show that the axial length l_p of the interpole (in cm.) and the flux density B in its air-gap (in kl./sq.cm.) are connected by the relation $Bl_p = 42.3$.

CHAPTER XIII

THE MECHANICAL FORCE AND TORQUE DUE TO ELECTROMAGNETIC ENERGY.

69. The Density of Energy in a Magnetic Field. The reader is already familiar with the fact that a certain amount of energy is required to establish the flux within a magnetic circuit, and that this energy remains stored in the field. This stored energy may be conveniently thought of as the kinetic energy of vortices around the lines of force (Art. 3). Various expressions for the total stored electro-magnetic energy are given in Arts. 57 and 58; the problem here is to find a relation between the distribution of the flux density and that of the energy in the field.

Consider first the simplest magnetic circuit (Fig. 1) consisting of a non-magnetic material. According to the last eq. (99), the total energy stored in such a circuit is

$$W = \tfrac{1}{2}\Phi^2 l/(\mu A) \text{ joules, } \quad \ldots \quad (162)$$

if Φ is in webers, l and A in cm., and $\mu = 1.257 \times 10^{-8}$ henrys per cm. cube. The volume of the field is $V = lA$ cubic cm. Since the flux density is uniform, the energy is also uniformly distributed, and the density of the energy is

$$W/V = \tfrac{1}{2}\Phi^2/(\mu A^2).$$

Denoting the density of the energy W/V by W', and introducing the flux density $B = \Phi/A$, we get

$$W' = \tfrac{1}{2}B^2/\mu \text{ joules per cu.cm. } \quad \ldots \quad (163)$$

Either B or μ can be eliminated from this expression by means of the relation $B = \mu H$, so that we have two other expressions for the density of the energy:

$$W' = \tfrac{1}{2}\mu H^2; \quad \ldots \ldots \ldots (164)$$

$$W' = \tfrac{1}{2}BH. \quad \ldots \ldots \ldots (165)$$

Two more expressions for the density of the energy can be written, using the reluctivity ν instead of the permeability μ.

In a uniform field the preceding expressions represent the actual amounts of energy stored per cubic centimeter. In a non-uniform field W' is the density of energy *at a point*, or the limit of the expression $\Delta W / \Delta V$. This is analogous to what we have in the case of a non-uniform distribution of matter, where the density of matter at a point is the limit of the ratio of the mass to the volume. Thus, the total energy stored in a non-uniform field is

$$W = \tfrac{1}{2}\nu \int_0^V B^2 dV, \quad \ldots \ldots \quad (166)$$

where the integration is to be extended over the volume of the whole magnetic circuit. Similarly, from eqs. (164) and (165) we get

$$W = \tfrac{1}{2}\mu \int_0^V H^2 dV, \quad \ldots \ldots \quad (167)$$

$$W = \tfrac{1}{2} \int_0^V HB dV. \quad \ldots \ldots \quad (168)$$

These expressions are consistent with eqs. (102) and (102a) as is shown in prob. 6 below.

When μ is variable, the preceding formulæ do not hold true, and the density of energy is represented by eq. (19), Art. 16.

Prob. 1. Deduce an expression for the magnetic energy stored in the insulation of a concentric cable (Fig. 46), between the radii a and b, the length of the cable being l cm. and the current i. Hint: For an infinitesimal shell of a radius x and thicknesss dx we have: $H = i/2\pi x$, and $dV = 2\pi x l \, dx$.   Ans. $W = 0.23 l i^2 \log (b/a) 10^{-8}$ joules.

Prob. 2. Check the answer to the preceding problem by means of eqs. (104) and (109).

Prob. 3. In a concentric cable (Fig. 46) $a = 7$ mm. and b is 20 mm. What is the density of the energy at the inner and outer conductors, when $i = 120$ amp.?

Ans.   4.68 and 0.57 microjoules per cu.cm.

Prob. 4. Deduce expression (110) from eq. (167).

Prob. 5. Taking the data from the various problems given in this book as typical, show that ordinarily in generators and motors a large proportion of the total energy of the field is stored in the air-gap.

Prob. 6. Show that eqs. (166) to (168) are consistent with eqs. (102) to (103a). Solution: Take an infinitesimal tube of partial linkages

(Fig. 45). The energy contained in this tube is $dW = \frac{1}{2}M_p d\Phi_p$; but $M_p = \int Hdl$, and $d\Phi = Bd.A$. Since $d\Phi$ is the same through all cross-sections of the tube, $d\Phi$ can be introduced under the integral sign, and we have $dW = \frac{1}{2}\int HdlBd.A = \frac{1}{2}\int HBdV$, the integration being extended over the volume of the tube. The total energy of the circuit is found by extending the integration over the volume of all the tubes of the field. The other equations are proved in a similar manner.

70. The Longitudinal Tension and the Lateral Compression in a Magnetic Field.

The existence of mechanical forces in a magnetic field is well known to the student. He needs only to be reminded of the supporting force of an electromagnet, of the attraction and repulsion between parallel conductors carrying electric currents, of the torque of an electric motor, etc. These mechanical forces must necessarily exist, if the magnetic field is the seat of stored energy. This is because, if we deform the circuit, we must in general change the stored energy and hence do mechanical work. The lines of force tend to shorten themselves and to spread laterally, so as to make the permeance of the field a maximum, with the complete linkages. Where there are partial linkages, it is the total stored energy that tends toward a maximum (Art. 57). This fact is entirely consistent with the hypothesis of whirling tubes of force, because the centrifugal force of rotation produces exactly the same effect, that is, a lateral spreading and a tension along the axis of rotation. A good analogy is afforded by a short piece of rubber tube filled with water and rotated about its longitudinal axis.

(a) *The Longitudinal Tension.* Consider again the simple magnetic circuit (Fig. 1), and let it be allowed to shrink, due to the longitudinal tension of the lines of force, so as to reduce its average length by Δl, without changing the cross-section A. Let at the same time the current be slightly decreased so as to keep the same total flux as before. Let F_l' be the mechanical tension along the lines of force, per square cm. of cross-section A; then the mechanical work done against the external forces which hold the winding stretched is $(F_l'.A)\,\Delta l$. The density of energy W' remains the same because B is the same, but the total stored energy is decreased by $W'(A\Delta l)$, because the volume of the field is decreased by $A\Delta l$. Since the change was made in such a way as to keep the total flux constant, no e.m.f. was induced in the winding during the deformation, and consequently there was no interchange of energy between the electric and the magnetic circuit. Thus, the decrease in the stored energy

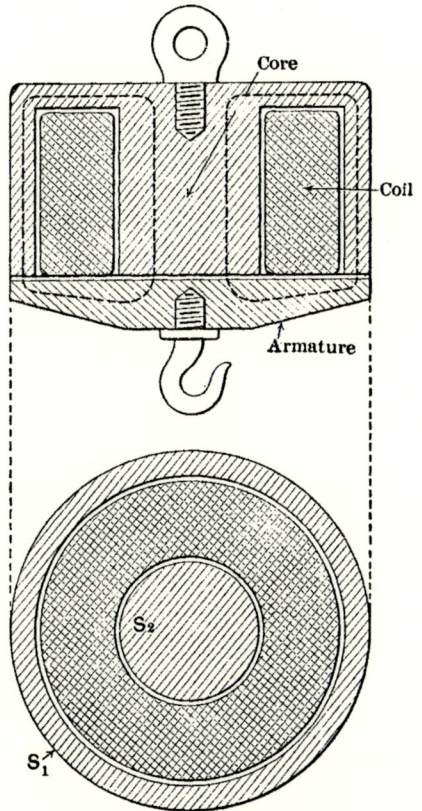

Fig. 58.—A lifting electromagnet.

is due entirely to the mechanical work performed. Equating the two preceding expressions, we have that

$$F_t' = W' = \tfrac{1}{2}B^2/\mu = \tfrac{1}{2}\mu H^2. \quad \ldots \quad (169)$$

If W' is in joules per cu.cm., F'_t is in *joulecens* per sq.cm. (see Appendix I), so that *in a rational system of units the mechanical stress per unit area is numerically equal to the density of the stored energy*. The physical dimensions of F' and W' are also the same.

If F_t' is in kg. per sq.cm., B in kilolines per sq.cm., and H in kiloampere-turns per cm., the preceding formula becomes, when applied to air,

$$F_t' = B^2/24.7 = H^2/15.6. \quad \ldots \quad (170)$$

These formulæ apply directly to the lifting magnet (Fig. 58), and give the carrying weight per unit area of the contact between the core and its armature. The total weight which the magnet is able to support is

$$F_t = AB^2/24.7 = AH^2/15.6 \text{ kg.}, \quad \ldots \quad (171)$$

where A is the sum of the areas denoted by S_1 and S_2. Of course, H is taken for the air-gap, which is the only part of the circuit that is changing its dimensions when the armature is moved.

(b) *The Lateral Compression*. Let now the simple magnetic circuit be allowed to expand laterally by a small length $\varDelta r$ in directions perpendicular to the surface of the toroid. Let F_c' be the pressure (compression) exerted by the lines of force upon the winding, per sq. cm. of the surface of the toroid. Then the mechanical work done by the magnetic forces in expanding the ring against the external forces which hold the winding, is $SF_c'\varDelta r$, where S is the surface of the toroid. Let again the current be slightly decreased during the deformation, so as to keep the flux constant. No voltage is induced in the winding, and hence there is no interchange of energy between the electric and the magnetic circuit. Thus we can find F_c', as we found the stress in the case of the tension, by equating the work done to the decrease in the stored energy. The stored energy is expressed by eq. (162), in which A is the only variable; hence by differentiating W with respect to A we get:

$$\varDelta W = -\tfrac{1}{2}\Phi^2 l\varDelta A/(\mu A^2) = -\tfrac{1}{2}B^2(l\varDelta A)/\mu.$$

This is a negative quantity, because the stored energy decreases. But $l\varDelta A$ represents the increase in the volume of the ring, so that $l\varDelta A = S\varDelta r$, and consequently

$$F_c' = \tfrac{1}{2}B^2/\mu = \tfrac{1}{2}\mu H^2 = W' = F_t'. \quad \ldots \quad (172)$$

In other words, *the lateral compression is numerically equal to the longitudinal tension, and both are numerically equal to the density of the stored energy.*

As an application of the lateral action, consider a constant-current or floating-coil transformer (Fig. 59), used in series arc-lighting. The leakage flux is similar in its character to that shown in Figs. 50 and 51. The lateral pressure of the leakage lines between the coils tends to separate them, acting against the weight of the floating coil. A part of this weight has to be balanced by a counter-weight Q because the electro-magnetic forces under normal operation are comparatively small.

Since the currents are alternating, the force is pulsating,

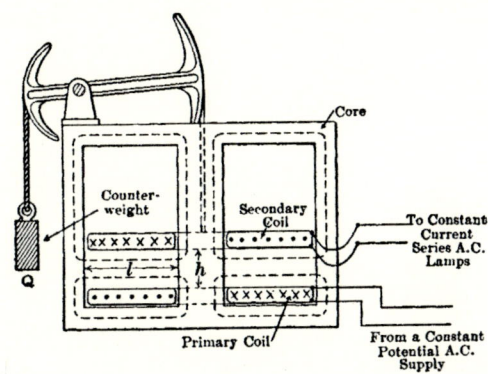

FIG. 59.—A floating-coil constant-current transformer.

but is always in the same direction, tending to separate the coils. The average force depends upon the average value of H^2, in other words, upon the effective value of the current. According to eqs. (170) and (172), we have

$$(F_c)_{ave} = (F_c')_{ave} \cdot S = H_{eff}^2 S/15.6 \text{ kg.,} \quad . \quad . \quad (173)$$

where S is the area of the floating coil in contact with the flux. With the assumed paths for the lines of force, and neglecting the reluctance of the iron core, we have that

$$H_{eff} = nI_{eff}/1000l \text{ kiloamp.-turns/cm.,} \quad . \quad . \quad (174)$$

where l is the length of the lines of force in the air, in cm., and nI_{eff} is the m.m.f. of either coil. The force of repulsion is proportional to the square of the current, and is independent of the distance h between the coils. Hence, a constant weight Q regulates for a constant current. When the coils are further from each other, the induced secondary voltage is less, on account of a much higher leakage flux. When the current increases momentarily, due to a decreasing line resistance, the coil is overbalanced and rises till the induced voltage and current fall to the proper value. Thus, the coil always floats at the proper height to induce the voltage needed on the line.

Formulæ (173) and (174) apply also to the mechanical forces between the primary and the secondary coils of a constant-potential transformer (Figs. 13 and 51). Under normal conditions these forces are negligible, but in a violent short-circuit the end-coils are sometimes bent away and damaged, unless they are properly secured to the rest of the winding. Such short-circuits are particularly detrimental in large transformers, having a close regulation, that is, having a very small internal impedance drop, and which are connected to systems of practically unlimited power and constant potential. As Dr. Steinmetz puts it, the closest approach to the appearance of such a transformer after a short-circuit is the way two express trains must look after a head-on collision at high speed.

Another interesting example of the effect of the mechanical forces produced by a magnetic field is the so-called *pinch phenomenon*.[1] The lines of force which surround a cylindrical conductor may be compared to rubber bands, which tend to compress it. With a liquid conductor and large currents, such for instance as are carried by a molten metal in some electro-metallurgical processes, the pressure of the magnetic field is sufficient to modify and to reduce the cross-section of the liquid conductor. This was first observed by Mr. Carl Hering and called by him the pinch phenomenon. In passing a relatively large alternating current through a non-electrolytic liquid conductor contained in a trough, he found that the liquid contracted in cross-section and flowed up-hill lengthwise in the trough, climbing up on the electrodes. With a further increase of

[1] E. F. Northrup, Some Newly Observed Manifestations of Forces in the Interior of an Electric Conductor, *Physical Review*, Vol. 24 (1907), p. 474. This article contains some cleverly devised experiments illustrating the pinch phenomenon, and also a mathematical theory of the forces which come into play.

current, this contraction of cross-section became so great at one point that a deep depression was formed in the liquid, with steeply inclined sides, like the letter V.

In most cases of mechanical forces in a magnetic field, these forces and the resulting movements are due to the combined

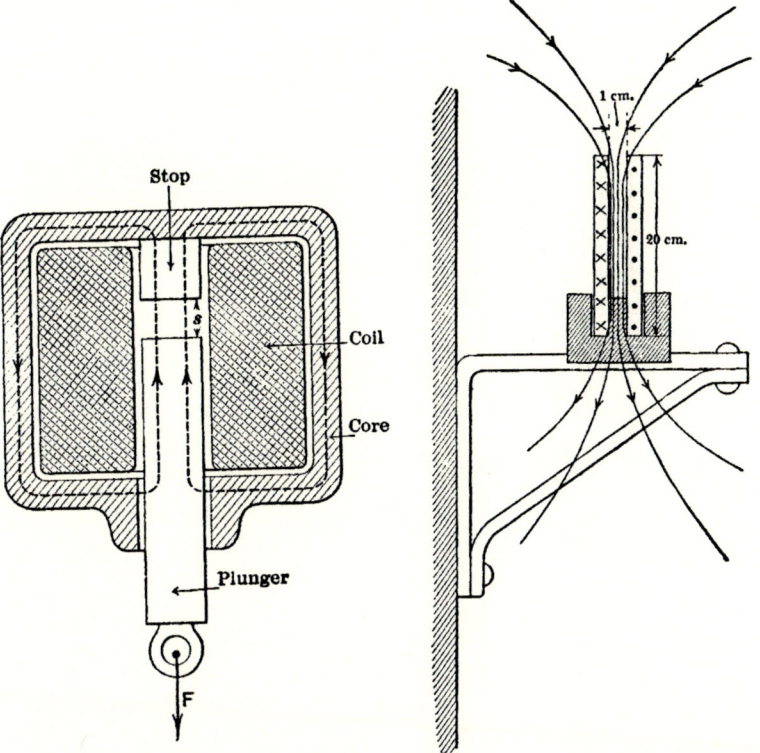

FIG. 60.—A tractive electromagnet.   FIG. 61.—Two bus-bars and their support.

action of longitudinal tensions and transverse compressions and not to one of these actions alone. For instance, a loop of flexible wire, through which a large current is flowing, tends to stretch itself so as to assume a maximum opening, that is, a maximum permeance of the magnetic field linked with it. This action is due to both the longitudinal tension and the lateral pressure. In such cases the mechanical forces are best computed by the principle of virtual displacements explained in the next article.

Prob. 7. Show that the required flux density in the air-gap of a lifting electromagnet (Fig. 58) can be calculated from the expression $B = 15.7\sqrt{\sigma F/A}$, in kl/sq.cm., where F is the rated supporting force, in metric tons, A is the area of contact in sq. dm., and σ is the factor of safety.

Prob. 8. Show that in an armored tractive magnet (Fig. 60) the tractive effort F varies with the air-gap s according to the law $Fs^2 = 3.08$ kg–cm.^{2}, when the excitation is 2000 amp.-turns and the cross-section of the plunger and of the stop is 12 sq.cm. Assume the leakage and the reluctance of the steel parts to be negligible.

Prob. 9. Referring to the preceding problem, what is the true average pull between the values $s = 1$ and $s = 4$ cm. and what is the arithmetical mean pull?   Ans.   0.77 and 1.63 kg.

Prob. 10. Indicate roughly the principal paths of magnetic leakage in Fig. 60, and explain the influence of the leakage upon the tractive effort, with a small and a large air-gap.

Prob. 11. The flux between two thin and high bus-bars, placed at a short distance from each other, has the general character shown in Fig. 61. Calculate the force per meter length that pushes the bus-bars apart when, during a short-circuit, the estimated current is 50 kilo-amperes.   Ans. About 800 kg. per meter.

Prob. 12. Deduce an expression for the magnetic pull due to the eccentric position of the armature in an electric machine (Fig. 62). A certain allowance is usually made for this pull in addition to the weight of the revolving part, in determining the safe size of the shaft.

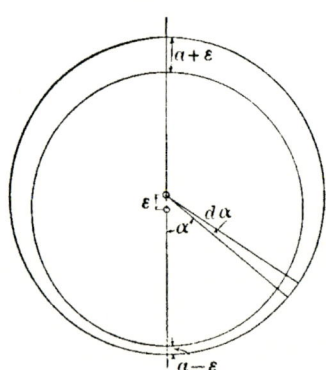

Solution: Since the pull is proportional to the square of the flux density, we replace the actual variable air-gap density by a constant radial density acting upon the whole periphery of the armature and equal to the quadratic average (the effective value) of the actual flux density distribution. Let this value be B_{eff} kl. per sq. cm. when the armature is properly centered. Let the original uniform air-gap be a, and the eccentricity be ε. Since a and ε are small as compared to the diameter of the armature, the actual air-gap at an angle α from the vertical is approximately equal to $a - \varepsilon \cos\alpha$. Neglecting the reluctance of the iron parts of the machine, the flux density is inversely as the length of the air-gap, so that we have

Fig. 62.—An eccentric armature.

$$B_\alpha = B_{eff}a/(a - \varepsilon \cos\alpha) = B_{eff}/[1 - (\varepsilon/a) \cos\alpha].$$

Let A be the total air-gap area to which B refers; then, according to eq. (171), the vertical component of the pull upon the strip of the width $d\alpha$ is

$$dF_{ex} = (B_a{}^2/24.7)(A/2\pi)d\alpha \cdot \cos \alpha.$$

The horizontal component of the pull is balanced by the corresponding component on the other half of the armature. The total pull downward is

$$F_{ex} = [2A/(2\pi \times 24.7)] \int_0^\pi B_a{}^2 \cos \alpha \; d\alpha.$$

Putting $\tan \tfrac{1}{2}\alpha = z$ and integrating we get the so-called Sumec formula for the eccentric pull, in kg.:

$$F_{ex} = (AB_{eff}{}^2/24.7)(\varepsilon/a)[1 - (\varepsilon/a)^2]^{-1.5}. \quad . \quad . \quad . \quad (175)$$

The integration is simplified if, before integrating, the difference is taken between the vertical forces at the points corresponding to α and to $\pi - \alpha$. The limits of integration are then 0 and $\tfrac{1}{2}\pi$.[1]

Prob. 13. The average flux density under the poles of a direct-current machine is 6.5 kl/sq.cm.; the poles cover 68 per cent of the periphery. The diameter of the armature is 1.52 m.; the effective length is 56 cm. What is the magnetic pull when the eccentricity is 10 per cent and when it is 50 per cent of the original air-gap?

Ans. 3.2 and 24 metric tons.

Prob. 14. The 22/2-kv., 2500-kva., transformer specified in prob. 5, Art. 64, had a total impedance drop of 73.5 volts at full load current on the low-tension side. What average force is exerted on each coil-face during a short-circuit, provided that the line voltage remains constant? The transformer winding consists of 12 high-tension coils of 100 turns each, and of 11 low-tension coils interposed between the high-tension coils, together with 2 half-coils at the ends. Two of the dimensions of the coils are repeated here: $O_m = 2.6$ m.; $l = 18$ cm. Hint: The short-circuit current is equal to $2000/73.5 = 27.2$ times the rated current.

Ans. About 22 metric tons.

Prob. 15. What is the mechanical pressure on the surface of the conductors in prob. 3?

Ans. 47.6 and 5.8 milligrams per sq.cm.

71. The Determination of the Mechanical Forces by Means of the Principle of Virtual Displacements.

In order to determine the mechanical force or torque between two parts of a magnetic circuit the general method consists in giving these parts an infinitesimal relative displacement and applying the law of

---

[1] J. K. Sumec, *Berechnung des einseitigen magnetischen Zuges bei Excentrizität*, *Zeitschrift für Elektrotechnik* (Vienna), Vol. 22 (1904), p. 727. This periodical is continued now under the name of *Electrotechnik und Maschinenbau*.

the conservation of energy to this displacement. From the equation so obtained the component of the force in the direction of the displacement can be calculated. Taking other displacements in different directions, a sufficient number of the components of the forces are determined to enable one to calculate the forces themselves. Since the forces in a given position of the system are perfectly definite, the result is the same no matter what displacements are assumed, provided that these displacements are *possible*, that is, consistent with the given conditions of the problem. Therefore displacements are selected which give the simplest formulæ for the energies involved. We have had two applications of this principle in the preceding article, in deriving the expressions for the tension and the compression in the field, by giving the simple magnetic circuit the proper "virtual" displacements. In applying this method, not only the mechanical displacement has to be specified, but also the electric and the magnetic conditions of the circuit, in order to make the energy relations entirely definite. Thus, in the preceding article, the electromagnetic condition was $\Phi = $ const.[1]

First let us take the case when the partial linkages are negligible; then according to the third eq. (99), the stored energy is

$$W_s = \tfrac{1}{2}\Phi^2 \mathcal{R}, \quad \ldots \quad \ldots \quad (176)$$

where $\mathcal{R}$ is the reluctance of the circuit. Let F be the unknown mechanical force between two parts of the magnetic circuit at a distance s, and let one part of the system be given an infinitesimal displacement ds. Let F be considered positive in the direction in which the displacement ds is positive. The mechanical work done is then equal to $F ds$. As in the preceding article, let this displacement take place with a constant flux, so that there is no interchange of energy between the magnetic circuit under consideration and the electric circuit by which it is excited. Then the work is done entirely at the expense of the stored energy of the magnetic circuit, and we have:

$$F ds = dW_m = -dW_s, \quad \ldots \quad \ldots \quad (177)$$

where dW_m is the mechanical work done. The sign minus before

[1] The principle of virtual displacements is much used nowadays in the theory of elasticity and in the calculation of the mechanical stresses in the so-called statically-indeterminate engineering structures.

dW_s is necessary because the stored energy decreases. From eqs. (176) and (177) we get

$$F = - \tfrac{1}{2}\Phi^2 . d\mathcal{R}/ds. \quad \ldots \ldots \quad (178)$$

In some cases it is more convenient to express F through M and $\mathcal{P}$. We have

$$F = -\tfrac{1}{2}\Phi^2 d(\mathcal{P}^{-1})/ds = +\tfrac{1}{2}(\Phi^2/\mathcal{P}^2)d\mathcal{P}/ds,$$

or

$$F = +\tfrac{1}{2}M^2 d\mathcal{P}/ds. \quad \ldots \ldots \quad (179)$$

In the preceding formulæ F is in joulecns, M is in ampere-turns, Φ is in webers, $\mathcal{P}$ is in henrys, and $\mathcal{R}$ in yrnehs. With other units the formulæ contain an additional numerical factor. It is to be noted that the mechanical forces are in such a direction that they tend to increase the permeance and decrease the reluctance of the circuit.   This agrees with previous statements. See Arts. 41 and 57.

If the partial linkages are of importance, it is convenient to express the stored energy in the form $W_s = \tfrac{1}{2}i^2 L$, because the inductance L takes account of the partial linkages; see eqs. (105) and (106), Art. 58. The energy equation, according to eq. (177), is then

$$Fds = -dW_s = -\tfrac{1}{2}d(i^2 L) \quad \ldots \ldots \quad (180)$$

and the condition that there is no interchange of energy with the line is

$$d(iL) = 0. \quad \ldots \ldots \quad (181)$$

The latter equation becomes clear by reference to eq. (106a), because $Li = n\Phi_{eq}$, where Φ_{eq} is the equivalent flux under the supposition of no partial linkages. The condition that there shall be no e.m.f. induced in the winding during the displacement, is $d(n\Phi_{eq})/dt = 0$, whence eq. (181) follows directly.

Performing the differentiations in eqs. (180) and (181), and substituting the value of Ldi from the second equation into the first, we get that

$$F = +\tfrac{1}{2}i^2 dL/ds. \quad \ldots \ldots \quad (182)$$

When there are no partial linkages, $L = n^2\mathcal{P}$, and eq. (182) becomes identical with (179).

Formulæ for the average force of direct current electromagnets. In a tractive magnet (Fig. 60) or a rotary magnet (Fig. 63) it is often required to know the *average* pull over a finite travel of the moving part. For the average force, the equation $F_{ave}\Delta s = \Delta W_m = -\Delta W_s$ holds, which is analogous to eq. (177) ; the only difference being that *finite* instead of *infinitesimal* increments are used. If the motion takes place at a constant flux, or at least the values of the flux are the same in the initial and the end-positions, we get from the preceding formulæ:

$$F_{ave} = \tfrac{1}{2}\Phi^2(\mathcal{R}_1 - \mathcal{R}_2)/(s_2 - s_1) ; \quad \ldots \quad (183)$$

$$= \tfrac{1}{2}(M_1^2\mathcal{P}_1 - M_2^2\mathcal{P}_2)/(s_2 - s_1) ; \quad . \quad . \quad (184)$$

$$= \tfrac{1}{2}(i_1^2 L_1 - i_2^2 L_2)/(s_2 - s_1). \quad . \quad . \quad (185)$$

The finite travel of the plunger often takes place at a constant current, for instance, in the regulating mechanism of a

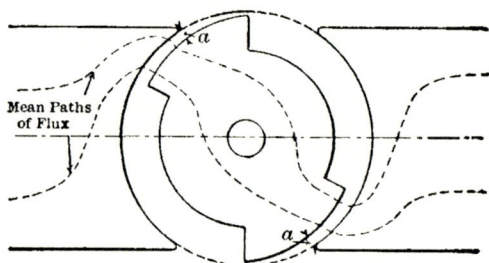

Mean Paths of Flux

Fig. 63.—A rotary electromagnet.

series arc-lamp, also approximately in a direct-current electromagnet connected across a constant-potential line. Under such conditions the foregoing formulæ are not directly appliable, because they have been deduced under the assumption of no interchange of energy between the electric and the magnetic circuits, so that this case has to be considered separately.

Let the motion be in the direction of the magnetic attraction, and let the current remain constant during the motion. The stored energy is larger in the end-position than in the initial position, because the flux is larger and the current is the same. Therefore, during the motion, the energy supplied from the line must be sufficient to perform the mechanical work, and to

increase the energy stored in the magnetic circuit.   The increase
in the stored energy is

$$\varDelta W_s = \tfrac{1}{2}i^2(L_2 - L_1),$$

and the energy supplied from the line is calculated as follows:
The average voltage induced during the motion of the plunger
is $e_{ave} = i(L_2 - L_1)/t$, where t is the duration of the motion.   The
energy supplied from the line is therefore

$$W_l = ie_{ave}t = i^2(L_2 - L_1).$$

Thus, the energy supplied from the line is twice as large as the
work performed, and we have the following important law
(due to Lord Kelvin): *When in a singly excited magnetic circuit
without saturation a deformation takes place at a constant current,
the energy supplied from the line is divided into two equal parts,
one half increasing the stored energy of the circuit, the other half
being converted into mechanical work.*

   According to this law we have, for a constant current electromagnet, that the mechanical work done is equal to the increase
in the energy stored in the magnetic field.   Hence

$$F_{ave}\varDelta s = \varDelta W_m = +\varDelta W_s.$$

Thus,

$$F_{ave} = \tfrac{1}{2}i^2(L_2 - L_1)/(s_2 - s_1); \quad \ldots \ldots \quad (186)$$

or, if the partial linkages are negligible,

$$F_{ave} = \tfrac{1}{2}(\varPhi_2{}^2\mathcal{R}_2 - \varPhi_1{}^2\mathcal{R}_1)/(s_2 - s_1); \quad \ldots \quad (187)$$

$$= \tfrac{1}{2}M^2(\mathcal{P}_2 - \mathcal{P}_1)/(s_2 - s_1). \quad \ldots \ldots \quad (188)$$

   When a magnet performs a rotary motion (Fig. 63), the
preceding formulæ are modified by substituting $Td\theta$ in place of
Fds, or $T_{ave}(\theta_2 - \theta_1)$ in place of $F_{ave}(s_2 - s_1)$.   Here T is the torque
in joules and $(\theta_2 - \theta_1)$ or $d\theta$ is the angular displacement of the
armature in radians.   Or else, in the foregoing formulæ F may
be understood to stand for the tangential force, and the displacement to be $ds = r \, d\theta$, where r is the radius upon which
the force F is acting.   Then the torque is $T = Fr$.   If, for instance,
we apply eq. (179) to a rotary motion, it becomes

$$T = Fr = +\tfrac{1}{2}M^2 d\mathcal{P}/d\theta. \quad \ldots \ldots \quad (189)$$

The other equations may be written by analogy with this one.

THE MAGNETIC CIRCUIT

Alternating-Current Electromagnets. The preceding formulæ are deduced under the supposition that the magnetic field is excited by a direct current. They are, however, applicable also to alternating-current electromagnets, because the pulsations in the current merely cause the energy to surge to and from the magnetic circuit, without any net effect, so far as the average stored energy and the mechanical work are concerned. The average stored energy corresponds to the effective values of the current and the flux.

In practice two types of alternating-current electromagnets are of importance, namely, those operating at a constant voltage and those operating at a constant current. As an example of the first class may be mentioned the electromagnets used for the operation of large switches at a distance (remote control); the windings of such electromagnets are usually connected directly across the line. Constant-current magnets are used in the operating mechanism of alternating-current series arc-lamps. In an A. C. electromagnet practically all of the voltage drop is reactive and hence proportional to the flux.

In a constant-potential electromagnet the effective value of the equivalent flux is the same for all positions of the plunger (neglecting the ohmic drop in the winding). Therefore, formula (185) holds true. Let e be the effective value of the constant voltage, or more accurately the reactive component alone, and let f be the frequency of the supply. Then, $e = 2\pi f L_1 i_1 = 2\pi f L_2 i_2$, so that the formula for the pull becomes

$$F_{a\,ve} = e(i_1 - i_2)/[4\pi f(s_2 - s_1)]. \quad . \quad . \quad . \quad (190)$$

For a constant-current A.C. electromagnet eq. (186) applies; introducing again the reactive volts $e_1 = 2\pi f L_1 i$ and $e_2 = 2\pi f L_2 i$, we get

$$F_{a\,ve} = i(e_2 - e_1)/[4\pi f(s_2 - s_1)]. \quad . \quad . \quad . \quad (191)$$

In both cases the mechanical work performed is proportional to the difference in the reactive volt-amperes consumed in the two extreme positions of the moving part.[1]

---

[1] For further details in regard to electromagnets consult C. P. Steinmetz, *Mechanical Forces in Magnetic Fields, Trans. Amer. Inst. Elec. Engs.*, Vol. 30 (1911), and the discussion following this paper; also C. R. Underhill *Solenoids, Electromagnets, and Electromagnetic Windings*(1910), chapters 6

Prob. 16. Derive formula (171) for the lifting magnet by means of the principle of virtual displacements. Solution: The reluctance of the air-gap is $\Re = s/(\mu A)$, so that $d\Re/ds = 1/(\mu A)$, hence, according to eq. (178) $F = -\frac{1}{2}\phi^2/(\mu A) = -AB^2/2\mu$. The minus sign indicates that the stress is one of tension.

Prob. 17. Derive expression (172) from formula (179).

Prob. 18. Derive the formula for the repulsion between the windings in a transformer from eq. (182).

Prob. 19. Derive from eq. (182) the force of repulsion between two infinitely long, parallel, cylindrical conductors placed at a distance of b meters apart, and forming an electric circuit (Fig. 47).

Ans. $2.04 i^2 (l/b) \times 10^{-8}$ kg. for l meters of the loop.

Prob. 20. What deformation of the windings may be expected during a severe short-circuit of a core-type or a cruciform type transformer (Figs. 12 and 14) with cylindrical coils, (a) when the centers of the coils are on the same horizontal line, and (b) when one of the windings is mounted somewhat higher than the other?

Prob. 21. Show that in a constant-current rotary magnet (Fig. 63) Ta = Const.—that is, the torque in the different positions of the armature is inversely proportional to the air-gap at the entering pole-tip. Hint: $d\mathcal{P} = \mu w r d\theta/a$, where w is the dimension parallel to the shaft.

Prob. 22. State Kelvin's law when mechanical work is done *against* the forces of the magnetic field.

Prob. 23. A 60-cycle, 8-amp., series arc-lamp magnet has a stroke of 32 mm.; the reactive voltage consumed in the initial position is 9 v., and in the final position 20 v. What is the average pull?

Ans. 372 grams.

72. The Torque in Generators and Motors.

The magnetic circuits considered in the preceding articles of this chapter are singly excited, that is, they have but one exciting electric circuit. From the point of view of mechanical forces this also applies to each air-gap in a transformer, because, neglecting the magnetizing current, the primary and the secondary coils may be combined into equivalent leakage coils (Art. 64). On the other hand, a generator or a motor under load has a *doubly-excited* magnetic circuit, the useful field being linked with both the field and the armature windings. The two m.m.fs. not being in direct opposition in space, the flux is deflected from the shortest

to 9 incl.; S. P. Thompson, On the Predetermination of Plunger Electromagnets, *Intern. Elect. Congress*, St. Louis, 1904, Vol. 1, p. 542; E. Jasse, Ueber Elektromagnete, *Elecktrotechnik und Maschinenbau*, Vol. 28 (1910), p. 833; R. Wikander, The Economical Design of Direct-current Electromagnets; *Trans. Amer. Inst. Elect. Engs.*, Vol. 30 (1911).

path, and the torque is due to the tendency of the tubes of force to shorten themselves longitudinally, and to spread laterally.

Consider the simplest generator or motor, consisting of a very long straight conductor which can move at right angles to a uniform magnetic field of a density B (Fig. 64). Let the ends of the conductor slide upon two stationary bars through which the current is conducted into a load circuit in the case of a generator action, and through which the power is supplied in the case of a motor action. If the magnetic circuit contains no iron, the resultant field is a superposition of the original uniform field and of the circular field created by the current in the conductor. With the direction of the current indicated in Fig. 64, the resultant field is stronger on the left-hand side of the conductor than it is on the right-hand side, and there is a resultant lateral pressure exerted upon the conductor to the right. In the case of a motor action the direction of the motion is in the same direction as the pressure from the stronger field. In the case of a generator action the conductor is moved by an external force against this pressure. Compare also the rule given in Art. 24.

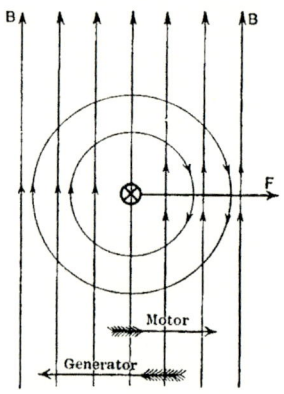

Fig. 64.—A straight conductor in a uniform magnetic field.

To find the mechanical force between the conductor and the field, we will apply again the principle of virtual displacements. Let the current through the conductor be i, and let the conductor be moved against the magnetic pressure by a small amount ds. Assume that the stored magnetic energy of the electric circuit to which the conductor belongs is the same in the various positions of the conductor. (Such is the case in actual machines.) Then the work done by the external force is entirely converted into electrical energy, and we have

$$F\,ds = ie\,dt, \quad . \quad . \quad . \quad . \quad . \quad . \quad (192)$$

where e is the induced e.m.f. Let both F and e refer to a length

l of the conductor.   Substituting the value of e from eq. (27), Art. 24, we have

$$F\,ds/dt = iBlv,$$

or, since $ds/dt = v$,

$$F = iBl. \quad . \quad . \quad . \quad . \quad . \quad . \quad (193)$$

In this expression i is in amperes, B is in webers per sq.cm., l is in cm., and F is in joulecens.   With other units the formula contains a numerical factor.

Formula (193) may be used also with a non-uniform field, and also when the direction of the conductor is not at right angles to that of the line of force.   In such cases the formula becomes $dF = iB\,dl$, where dF is the force acting upon an infinitesimal length dl of the conductor, and B is understood to be the component of the actual flux density perpendicular to dl.

As an application of formula (193), consider the attraction or the repulsion between two straight parallel conductors carrying currents i_1 and i_2, and placed at a distance b from each other. The circuit of each conductor may be considered closed through a concentric cylindrical shell of infinite radius, as in Art. 60. It is apparent from symmetry that the field produced by each system gives no resultant force with the current in the same system.   Thus, the mechanical force is due to the action of the field 1 upon the current 2, and *vice versa*.

The flux density due to the system 2 at a distance b from the conductor 2 is $B = \mu . i_2/2\pi b$, so that, according to eq. (193),

$$F = \mu i_1 i_2 l/2\pi b \text{ joulecens}, \quad . \quad . \quad . \quad . \quad (194)$$

where $\mu = 1.257 \times 10^{-8}$.   In kilograms the same formula is

$$F = 2.04 i_1 i_2 (l/b) 10^{-8}, \quad . \quad . \quad . \quad . \quad (194a)$$

provided that l and b are measured in the same units.   The force is an attraction or a repulsion according to whether the two currents are flowing in the same or in the opposite directions. When $i_1 = i_2$, this formula checks with that given in prob. 19 above.

Formula (193) applies also to the tangential force between the field and armature conductor in any ordinary generator or motor, provided that (a) the conductors are placed upon a smooth-body armature, and (b) the conductors are distributed

uniformly over the armature periphery, so that the stored magnetic energy is the same in all positions of the armature. It
would be entirely wrong, however, to apply this formula to
a slotted armature, using for B the actual small flux density
in the slot within which the conductor lies. This would give
the force acting upon the conductor itself, and tending to press
it against the adjacent conductor or against the side of the slot;
but the actual tangential force exerted upon the armature as
a whole is many times greater, and practically all of it is exerted
directly upon the steel laminations of the teeth.

At no-load, the flux distribution in the active layer of the
machine is symmetrical with respect to the center line of each
pole (Fig. 24), so that the resultant pull along the lines of force
is directed radially. The armature currents distort the field
as a whole, and also distort it locally around each tooth, the
general character of distortion being shown in Fig. 36. The
unbalanced pull along the lines of force has a tangential component which produces the armature torque. This torque,
although caused by the current in the armature conductors,
is largely exerted directly upon the teeth, because the flux density
there is much higher.

Thus, in order to determine the total electromagnetic torque
in a slotted armature, it is again necessary to apply the principle
of virtual displacements. The reluctance of the active layer
per pole varies somewhat with the position of the armature,
so that the energy stored in the field is also slightly fluctuating.
It is convenient, therefore, to take a displacement which is a
multiple of the tooth pitch, in order to have the same stored
energy in the two extreme positions. This gives the average
electromagnetic torque.

(a) *The Torque in a Direct-current Machine.* Let the virtual
displacement be equal to θ geometric degrees and be accomplished
in t seconds. Then we have

$$T_{ave}\theta = iEt, \quad \ldots \ldots \ldots \quad (195)$$

where T is the torque, i is the total armature current, and E is
the total induced e.m.f. Eq. (195) states the equality of the
mechanical work done and of the corresponding electrical energy
supplied. The average induced e.m.f. is independent of the
flux distribution, or of the presence or absence of teeth (see

Art. 24 and prob. 18 in Art. 26). Take θ to correspond to two pole pitches, or $\theta = 2\pi/(\frac{1}{2}p)$; then $t = 1/f$, where f is the frequency of the magnetic cycles. Substituting these values and using the value of E from eq. (37), Art. 31, we get, after reduction,

$$T_{ave} = iNp\Phi/\pi \text{ joules,} \quad \ldots \ldots \quad (196)$$

where Φ is in webers; or

$$T_{ave} = 0.0325iNp\Phi \times 10^{-2} \text{ kg-meters,} \quad \ldots \quad (196a)$$

Φ being in megalines.

This formula does not contain the speed of the machine, the torque depending only upon the armature ampere-turns iN and the total flux $p\Phi$. Consequently, the formula can be used for calculating the starting torque or the starting current of a motor. Eqs. (196) and (196a) give the total electromagnetic torque, part of which serves to overcome the hysteresis, eddy currents, friction and windage. The remainder is available on the shaft. When calculating the starting torque, it is necessary to take into account the effort required for accelerating the revolving masses.

(b) *The Torque in a Synchronous Machine.* The equation of energy is

$$T_{ave}\theta = miE \cos \phi'.t, \quad \ldots \ldots \quad (197)$$

where m is the number of phases, i and E are the effective values of the armature current and the induced voltage per phase, and ϕ' is the internal phase angle (Fig. 37). Taking again a displacement over two poles and using the value of E from eq. (3) Art. 26, we get

$$T_{ave} = 0.0361k_bmiN \cos \phi'p\Phi10^{-2} \text{ kg-meters.} \quad . \quad (198)$$

(c) *The Torque in an Induction Machine.* The torque being exerted between the primary and the secondary members of an induction machine, it may be considered from the point of view of either member. This is because the torque of reaction upon the stator is equal and opposite to the direct torque upon the rotor. For purposes of computation it is more convenient to consider the torque from the point of view of the primary winding, in order to be able to use the primary frequency and

the synchronous speed.   Therefore eq. (198) gives the torque of
an induction machine (including, as before, friction and hysteresis),
where the various quantities refer to the stator.   However, these
quantities may equally well be taken in the secondary, but in
this case, since hysteresis occurs mainly in the stator, the torque
to overcome hysteresis is not included.

Formula (198) is hardly ever used in practice, especially
for the computation of the starting torque, because it is difficult
to eliminate the large leakage flux which gives no torque.   It
is much more convenient to determine the torque from the circle
diagram, or from the equivalent electric circuit.

In case the torque is determined from the equivalent electric circuit, we can write from eqs. (195) and (197) the expressions for the
torque directly, by substituting for θ its value $2\pi \cdot t \cdot (\text{R.P.M.})/60$.
For a direct-current machine,

$$T_{ave} = 0.0325 \frac{30iE}{\pi \cdot (\text{R.P.M.})} \text{Kg.-meters.} \quad . \quad . \quad . \quad (199)$$

Here the induced e.m.f. $E = E_t \pm iR_a$, according to whether the
machine is a generator or a motor; E_t is the terminal voltage
and R_a the resistance of the armature, brushes, and series field.

For an alternating-current machine,

$$T_{ave} = 0.0325 \frac{30miE \cos \phi'}{\pi (\text{R.P.M.})} \text{Kg.-meters.} \quad . \quad . \quad . \quad (200)$$

In a synchronous machine E is the induced e.m.f. and ϕ' is
the internal phase angle.   In an induction machine, $iE \cos \phi'$
is the power per phase delivered to the rotor, that is, the input
minus hysteresis and primary I^2R loss.   The term (R.P.M.) is
in all cases the synchronous speed of the machine.

Prob. 24.   Two single-conductor cables from a direct-current machine
are installed parallel to each other at a distance of 16 cm. between their
centers, on transverse supports spaced 80 cm. apart.   The rated current
through the cables is 850 amp.   What is the force acting upon each
support under the normal conditions, and when the current rises to
twenty times its rated value during a short-circuit?
Ans.   0.0737 and 29.5 kg.

Prob. 25.   A 4-pole, series direct-current motor must develop a
starting torque of 74 kg.-m. (including the losses).   The largest possible
flux per pole is about 2.5 ml.; there are 240 turns in series between the
brushes.   Calculate the starting current.   Ans.   95 amp.

Prob. 26. Explain the reason for which the compensating winding (Art. 54), while removing the armature reaction, does not affect appreciably the useful torque of the motor. Hint: this can be shown by applying the method of virtual displacements; also from the fact that the local distortion of the flux in the teeth is not removed.

Prob. 27. On the basis of Art. 15, explain the mechanism by which the phenomenon of hysteresis in an armature core causes an opposing torque. Explain the same for eddy currents.

Prob. 28. Demonstrate that formula (193) may be used for slotted armatures, provided that B stands for the average flux density per tooth pitch. Hint: take a virtual displacement of one tooth pitch, and express the induced voltage through the flux $\Delta\Phi = Bl\lambda$ per tooth pitch.

Prob. 29. Show that in a single-phase synchronous motor the torque at the synchronous speed pulsates at double the frequency of the supply; also that the torque is zero at any but the synchronous speed.

Prob. 30. Prove that in an induction machine the torque near synchronism is approximately proportional to the square of the voltage and to the per cent slip. Hint: $i = \text{Const.} \times \Phi \times \text{slip}$.

Prob. 31. Describe in detail how to calculate the maximum starting torque of a given induction motor, and how to calculate the amount of secondary resistance necessary for a prescribed torque. See the author's *Electric Circuit*, Chap. XII.

APPENDIX I

THE AMPERE-OHM SYSTEM OF UNITS

THE ampere and the ohm can be now considered as two arbitrary fundamental units established by an international agreement. Their values can be reproduced to a fraction of a per cent according to detailed specifications adopted by practically all civilized nations. These two units, together with the centimeter and the second, permit the determination of the values of all other electric and magnetic quantities. The units of mass and of temperature do not enter explicitly into the formulæ, but are contained in the legal definition of the ampere and of the ohm. The dimension of resistance can be expressed through those of power and current, according to the equation $P = I^2 R$, but it is more convenient to consider the dimension of R as fundamental in order to avoid the explicit use of the dimension of mass [M].

For the engineer there is no more a need of using the electrostatic or the electromagnetic units; for him there is but one *ampere-ohm system*, which is neither electrostatic nor electromagnetic. The ampere has not only a magnitude, but a *physical* dimension as well, a dimension which with our present knowledge is fundamental, that is, it cannot be reduced to a combination of the dimensions of length, time, and mass (or energy). Let the dimension of current be denoted by [I] and that of resistance by [R]; let the dimensions of length and time be denoted by the commonly recognized symbols [L] and [T]. The magnitudes and the dimensions of the important magnetic units are expressed through these four, as is shown in the following table. For the expressions of the electric and the electrostatic quantities in the ampere-ohm system see Appendix to the author's *Electric Circuit*

Other units of more convenient magnitude are easily created by multiplying the above-tabulated units by powers of 10, or by adding prefixes milli-, micro-, kilo-, mega-, etc.

A study of the physical dimensions of the magnetic quantities is interesting in itself, and gives a better insight into the nature of these quantities. Moreover, formulæ can sometimes be checked by comparing the physical dimensions on both sides of the equation. Let, for instance, a formula for energy be given

$$W = k\Phi B \nu l,$$

where k is a numerical coefficient. Substituting the physical dimensions of all the quantities on the right-hand side of the equation from the table below, the result will be found to have the dimension of energy. This fact adds to one's assurance that the given formula is theoretically correct.

TABLE OF MAGNETIC UNITS AND THEIR DIMENSIONS IN THE AMPERE-OHM SYSTEM

Symbol and Formula	Quantity.	Dimension.	Name of the Unit.
$M = nI$	Magnetomotive force	$[I]$	Ampere-turn.
$H = M/l$	Field intensity, or m.m.f. gradient	$[IL^{-1}]$	Ampere-turn per centimeter.
$\Phi = ET$	Magnetic flux	$[IRT]^*$	Weber (maxwell).
$B = \Phi/A$	Magnetic flux density	$[IRTL^{-2}]$	Weber (maxwell) per square centimeter.
$\mathcal{P} = \Phi/M$	Permeance	$[RT]$	Henry (perm).
$\mathcal{R} = M/\Phi = 1/\mathcal{P}$	Reluctance	$[R^{-1}T^{-1}]$	Yrneh (rel).
$\mu = B/H$	Permeability	$[RTL^{-1}]$	Henry (perm) per centimeter cube.
$\nu = H/B = 1/\mu$	Reluctivity	$[R^{-1}T^{-1}L]$	Yrneh (rel) per centimeter cube.
$W = \frac{1}{2}M\Phi$	Magnetic energy or work	$[I^2RT]$	Joule or watt-second.
$W/v = \frac{1}{2}BH$	Density of magnetic energy	$[I^2RTL^{-3}]$	Joule per cubic centimeter.
$F = W/l$	Force	$I^2RTL^{-1}]$	Joulecen.

* This is also the dimension of the magnetic pole strength. The concept of pole strength is of no use in electrical engineering, and, in the author's opinion, its usefulness in physics is more than doubtful. The whole theory of electromagnetic phenomena can and ought to be built up on the two laws of circuitation, as has been done by Oliver Heaviside in his *Electromagnetic Theory*.

A small irregularity is due to the use of the maxwell and of its multiples instead of the weber. As long as this usage persists it is convenient to use the corresponding units for reluctance and permeance, to which the author has ventured to give the names of rel and perm. Since one maxwell is equal to $1/10^8$ of a weber, one perm is equal to $1/10^8$ of one henry, and one rel is 10^8 yrnehs. Accordingly, permeabilities and reluctivities are measured in perms per centimeter cube and in rels per centimeter cube respectively.

In order not to break with the established usage, the maxwell, the perm, and their multiples are employed in numerical computations in this book, while the weber and the henry are used in the deduction of the formulæ, being the natural fundamental units of flux and permeance in the ampere-ohm system. It is possible that the constant necessity for multiplying or dividing results by 10^{-8}, due to the use of the maxwell, may prove to be more and more of an inconvenience in proportion as magnetic computations come into common engineering practice. Then the weber, the henry, and their submultiples will be found ready for use, and the system of magnetic units will be completely coordinated.

Another irregularity in the system as outlined above is caused by the use of the kilogram as the unit of force, because it leads to two units for energy and torque, viz., the kilogram-meter and the joule; 1 kg.-meter = 9.806 joules. *Force ought to be measured in joules per centimeter length*, to avoid the odd multiplier. Such a unit is equal to about 10.2 kg., and could be properly called the *joulecen* ($= 10^7$ dynes). There is not much prospect in sight of introducing this unit of force into practice, because the kilogram is too well established in common use. The next best thing to do is to derive formulæ and perform calculations, whenever convenient, in joulecens, and to convert the result into kilograms by multiplying it by 10.2. This is done in some places in this book.

Thus, leaving aside all historical precedents and justifications, the whole system of electric and magnetic units is reduced to this simple scheme: In addition to the centimeter, the gram, the second and the degree Centigrade, two other fundamental units are recognized, the ohm and the ampere. All other electric and magnetic units have dimensions and values which are connected with those of the fundamental six in a simple and almost self-evident manner (see the table above).

To appreciate fully the advantages of the practical ampere-ohm system over the C.G.S. electrostatic and electromagnetic systems, one has only to compare the dimensions, for instance, of magnetomotive force and of flux in these three systems, as shown below.

	The Ampere-Ohm System.	C.G.S. Electromagnetic System.	C.G.S. Electrostatic System.
Dimension of m.m.f.	[I]	$L^{\frac{1}{2}}M^{\frac{1}{2}}T^{-1}\mu^{-\frac{1}{2}}$	$L^{\frac{3}{2}}M^{\frac{1}{2}}T^{-2}\kappa^{\frac{1}{2}}$
Dimension of flux.	[IRT]	$L^{\frac{3}{2}}M^{\frac{1}{2}}T^{-1}\mu^{\frac{1}{2}}$	$L^{\frac{1}{2}}M^{\frac{1}{2}}\kappa^{-\frac{1}{2}}$

APPENDIX II

AMPERE-TURN vs. GILBERT

THE reader has probably been taught before that the permeability of air is equal to unity in the electromagnetic C.G.S. system; silent assumption was then probably made that $\mu = 1$ also in the practical ampere-ohm system. The true situation is, however, as follows: In any system of units whatsoever, the fundamental equation $\Phi = (\mu A/l) \cdot M$ holds true, being a mathematical expression of an observed fact. Now let the quantities be expressed in the ampere-ohm system, and assume the centimeter to be the unit of length. The flux is then expressed either in maxwells or in webers, both of which are connected with the ampere-ohm system through the volt. *The natural (though not the only possible) unit for the magnetomotive force is one ampere-turn.* Therefore, all the quantities in the foregoing equation are determinate, and the value of μ cannot be prescribed or assumed, but must be determined from an actual experiment, the same as the electric conductivity of a metal, or the permittivity of a dielectric have to be determined. Experiment shows that $\mu = 1.257$ when the maxwell is used as the unit of flux, and hence $\mu = 1.257 \times 10^{-8}$ if the flux is measured in webers.

It is possible to assume $\mu = 1$, provided that the unit of magnetomotive force is not prescribed in advance. In this case, the unit of magnetomotive force, as determined from experiment, comes out equal to $1/1.257$ of an ampere-turn. This unit is called the *gilbert,* and it must be understood that the permeability of non-magnetic materials is equal to unity only if the magnetomotive force is measured in gilberts. To the author the advantages of such a system for practical use are more than doubtful. In the first place, the gilbert is a superfluous unit, because the results of calculations must after all *for practical purposes* be converted into ampere-turns in order to specify the number of turns and the exciting current of windings. Thus, one would have to deal with two units of magnetomotive force, the gilbert and the ampere-turn, one being about 0.8 of the other. In the second place, with the assumption $\mu = 1$ for non-magnetic materials B becomes numerically equal to H, which is a grave inconvenience, because B and H are different physical quantities. B and H have different physical dimensions, because μ has a definite physical dimension, even though the numerical value of it is assumed to be equal to unity for air. Therefore, to be sure that proper physical dimensions are preserved, one has to remember where μ is omitted in formulæ, and for a physical interpretation of results it is much more convenient to have it there, explicitly.

Still another objection to using the gilbert and to putting μ equal to unity for air is that the ratio of the ampere-turn to the gilbert is equal to a quasi-scientific constant $4\pi/10$. To the author's knowledge, there is no simple, elementary way of deducing the value of this constant, without going over the whole mathematical theory of electricity and magnetism. Thus, a constant is retained in practical formulæ, the significance of which remains a puzzle to the engineer all his life. It is true that the value of $\mu = 1.257$ is equal to the same $4\pi/10$ after all; but in this case there is nothing "absolute," mysterious, or sacred about the value of $4\pi/10$. The student is simply told that 1.257 happens to be equal to $4\pi/10$ because the value of the ampere was unfortunately so selected. It is not necessary to go into further details, because the historical reasons which led to the selection of the values of unit pole and unit current hardly hold at present. All calculations would be just as convenient if μ were equal to 2.257, or any other value, instead of 1.257.

For these reasons the author unhesitatingly discards the gilbert in teaching as well as in practice and uses the ampere-turn as the natural unit of magnetomotive force. The value of permeability becomes then an experimental quantity which depends upon the units selected for flux and length.

APPENDIX III

THE SQUIRREL-CAGE ROTOR

(To Supplement Pages 134 to 136)

Equations (65) to (70) are deduced for an induction motor with a phase-wound rotor, and one must be careful in applying them to a squirrel-cage rotor. In such a rotor there is one bar per phase per pair of poles, or one quarter of a turn per pole per phase. Consequently, if C_2 is the total number of secondary bars, and p is the number of poles, we must substitute in eq. (65)

$$m_2 = C_2/(\tfrac{1}{2}p); \quad k_{b2} = 1, \quad \text{and} \quad n_2 = 0.25.$$

This gives, after reduction,

$$\frac{\text{equivalent primary current per phase}}{\text{secondary current per bar}} = \frac{C_2}{2pn_1m_1k_{b1}} = \frac{C_2}{k_{b1}C_1}, \quad (65a)$$

where $2pn_1m_1 = C_1$ is the total number of primary conductors. Since there are $\tfrac{1}{2}p$ rotor bars in parallel, the total secondary current per phase is equal to $\tfrac{1}{2}p$ times the current per bar. Eq. (65a) may also be deduced directly from the fundamental eq. (64); the m.m.fs of the actual rotor and of the equivalent rotor must be equal, so that $k_{b1}C_1i_2' = C_2i_b$, or $i_2'/i_b = C_2/(k_{b1}C_1)$.

In eq. (66), if e_2' and e_2 be understood as the voltages induced per pole per phase, e_2 is the voltage induced in one-half of one bar, because there is one bar per pair of poles per phase. In this case n_2 is again equal to 0.25. It is more convenient in some cases to take e_2 as the voltage per bar. Then e_2' is the equivalent voltage *per pair of poles;* n_1 is the number of primary turns per phase per pair of poles; and $n_2 = 0.50$, there being one bar, or one-half of a turn, per pair of poles per phase. One may also take e_2'/e_2 to represent the ratio of voltages per phase, including all the poles. In this case n_1 is the total number of primary turns in series per phase, and $n_2 = 0.50$ as before, because all the secondary bars are connected in parallel. Thus,

$$\frac{\text{equivalent voltage per primary phase}}{\text{secondary voltage per bar}} = \frac{k_{b1}pn_1}{0.5} = \frac{k_{b1}C_1}{m_1}. \quad (66a)$$

In eqs. (67) and (68), if r_2 and r_2' be understood as the resistances per pole per phase, r_2 represents that of one-half of a bar, augmented of course for the influence of the end-rings. In this case, as before, $m_2 = C_2/(\tfrac{1}{2}p)$, and $n_2 = 0.25$. If, however, it is desired to understand under r_2 the resistance of one whole bar, r_2' is the equivalent primary resistance *per pair of poles*. In some cases it is convenient to know the ratio between the equivalent resistance R_2' per phase of the primary circuit and the resistance R_b of one secondary bar (with the resistance of the end-rings taken into account). Eq. (67) becomes

$$C_2 i_b{}^2 R_b = m_1 i_2'^2 R_2' \quad \ldots \ldots \quad (67a)$$

Substituting the ratio of the currents from eq. (65a) we find that

$$\frac{\text{equivalent resistance per primary phase}}{\text{resistance per secondary bar}} = \frac{(k_{b1}C_1)^2}{m_1 C_2}. \quad (68a)$$

A similar relation holds for the ratio of inductances or reactances.

INDEX

PAGE

280 INDEX

284

Printed in the United States
27889LVS00002BA/6